SpringerBriefs in Mathematics

SpringerBriefs in Mathematics showcases expositions in all areas of mathematics and applied mathematics. Manuscripts presenting new results or a single new result in a classical field, new field, or an emerging topic, applications, or bridges between new results and already published works, are encouraged. The series is intended for mathematicians and applied mathematicians.

For further volumes:
http://www.springer.com/series/10030

Călin-Ioan Gheorghiu

Spectral Methods for Non-Standard Eigenvalue Problems

Fluid and Structural Mechanics and Beyond

 Springer

Călin-Ioan Gheorghiu
T. Popoviciu Institute of Numerical
  Analysis
Romanian Academy
Cluj-Napoca
Romania

ISSN 2191-8198   ISSN 2191-8201   (electronic)
ISBN 978-3-319-06229-7   ISBN 978-3-319-06230-3   (eBook)
DOI 10.1007/978-3-319-06230-3
Springer Cham Heidelberg New York Dordrecht London

Library of Congress Control Number: 2014936233

Mathematical Subject Classification (2010): 34B09, 34B15, 34B16, 34B40, 65F15, 65L11, 65L15, 65L60, 74S25, 76E05, 76M22

Printed on acid-free paper

Springer is part of Springer Science+Business Media (www.springer.com)

Preface

The following empirical rule is sound advice to number-crunchers and arithmurgists: *Parabola/Sine Rule*. If a solution is well-approximated by a parabola or a sine function or otherwise is very, very smooth, then it is a poor example for comparing and evaluating numerical methods because even a dreadful, awful numerical method will work tolerantly well for such nice function.

—J. P. Boyd, 2011

Spectral methods, *collocation*, *Galerkin* and *tau*, offer useful alternatives to finite differences or finite elements type methods in solving boundary value problems attached to differential equations. Advantages of such methods include the production of global solutions which are rapid convergent and, in some cases, the avoidance of Gibbs phenomenon at domain boundaries.

This work is not oriented on formal reasoning which means the well known sequence of axioms, theorems, proofs, corollaries, etc. Instead, it is mainly oriented to the *constructive and practical aspects* of spectral methods. Consequently, we rigorously examine the most important qualities as well as drawbacks of spectral methods in the context of numerical methods devoted to solve nonstandard eigenvalue problems. Some nonlinear singularly perturbed boundary value problems along with eigenproblems obtained by their linearization around constant solutions are also considered.

By non-standard eigenvalue problems, we mean singular Sturm–Liouville problems, high order (larger than two) singular and nonsingular generalized eigenvalue problems, eigenproblems involving boundary conditions dependent on the eigenparameter and multiparameter eigenvalue problems. We consider them challenging and thus suitable in evaluating spectral methods, according to above Boyd's advice. Thus, one of the main aims of this work is to review our contributions in tuning spectral methods in order to obtain reliable eigenvalues and eigenmodes, at least for a specified region of the spectrum.

For problems formulated on finite domains, we have used mainly families of Chebyshev polynomials in order to build up test and trial spaces for all three methods. In working with spectral tau and Galerkin methods, we have succeeded to construct both type of these fundamental spaces such that the discretization matrices inherit the properties of differential operators and additionally are sparse (even banded) with fairly good conditioning properties.

With respect to the Chebyshev collocation, its spectral accuracy has enabled us, among other results, to formulate an important conjecture corresponding to the first eigenvalue of the singularly perturbed Viola's problem.

For problems defined on the half-line, we have considered only collocation methods based on Laguerre functions. With such basis functions, we avoid the domain truncation or mapping and enforce exactly any type of boundary conditions at infinity.

Frequently, tau scheme or collocation coupled with some factorization of a high-order differential operator cast a problem into singular algebraic generalized eigenvalue problems. They are singular in the sense that the first matrix in the pencil has a larger rank than the second one. In this situation, some spurious eigenvalues (at infinity) are inevitable when QR/QZ algorithms are used. Moreover, for high order eigenvalue problems, i.e., sixth and eighth orders, and for orders of thousands for the cut-off parameters, the above-mentioned collocation algorithms could be expensive with respect to the CPU consumed time and storage. Thus, in order to improve the accuracy in computing a specified part of the finite spectrum and to shorten the elapsed time we adapted some subspace methods to solve eigenproblems such as Jacobi–Davidson methods. These methods are target-oriented and systematically avoid spurious eigenvalues.

Two classes of applications from mechanics of continua are envisaged. In the first one, the linear stability of elastic systems along with some linear hydrodynamic stability problems are analyzed. Both type of problems lead, in some instances, to eigenvalue problems containing eigenparameter-dependent boundary conditions. In the second class, we gather a lot of second and fourth order genuinely nonlinear two-point boundary value problems formulated on finite or infinite intervals. Some of them exhibit singularities in origin and at infinity and are originated in fluid mechanics, foundation engineering, etc.

The work can be used as a self contained supplementary textbook for various review courses. One has all ingredients, i.e., differentiation matrices, in both physical and phase spaces, and complete procedures in order to implement various types of boundary conditions. Consequently, linear and nonlinear two-point boundary value problems, possibly singularly perturbed, as well as eigenvalue problems of various orders can be solved.

I have to acknowledge first one special colleague who worked with enthusiasm besides me in spectral methods early in the turbulent years '90. He is Sorin Iuliu Pop (TU Eindhoven and alumnus of UBB Cluj-Napoca). I am lucky to have found Sorin at a stage of his career when he still had so much time to spend with others. I am especially grateful to Bor Plestenjak (University of Ljubljana) for providing suggestive colorful pictures concerning Mathieu's eigenmodes. To him, Joost Rommes (NXP Semiconductors, Eindhoven) and Michiel Hochstenbach (TU Eindhoven) must go my thanks for many illuminating discussions on the implementation of JD methods.

This book could not have come to be without a lot of kind assistance from Springer. I have made great profit from the remarks, comments and suggestions of all referees.

Cluj-Napoca, Romania, February 2014   Călin-Ioan Gheorghiu

Contents

Acronyms

CGauss	Chebyshev–Gauss (nodes) quadrature formula
CGaussL	Chebyshev–Gauss–Lobatto (nodes) quadrature formula
CGaussR	Chebyshev–Gauss–Radau (nodes) quadrature formula
ChC	Chebyshev collocation method
ChG	Chebyshev Galerkin method
ChGHS	Chebyshev–Galerkin method based on Heinrichs' basis (trial) and Shen's basis (test)
ChGS	Chebyshev–Galerkin method based on Shen's basis as test and trial functions
ChT	Chebyshev tau method
$D^{(2)}$	Second order Chebyshev differentiation matrix on (CGaussL) nodes
$\widetilde{D}^{(2)}$	Second order Chebyshev differentiation matrix on $\mathbf{x}_{int}$, i.e., the Dirichlet homogeneous boundary conditions are enforced
EHD	Electrohydrodynamic
GD^k	Galerkin differentiation matrix of order k
GMRES	Generalized minimal residuals method
I	The unitary matrix of order $N-1$
IRAM	Implicitly restarted Arnoldi method
JD	Jacobi–Davidson method
LC	Laguerre collocation method
$LD^{(k)}$	The k order Laguerre differentiation matrix on the roots of Laguerre polynomial of order $N-1$ with the origin added
$\widetilde{LD}^{(k)}$	The k order Laguerre differentiation matrix on the roots of Laguerre polynomial of order $N-1$, i.e., a homogeneous boundary condition is enforced in origin
$\widetilde{\widetilde{LD}}^{(k)}$	The k order Laguerre differentiation matrix on the roots of Laguerre polynomial of order $N-1$, with two homogeneous boundary conditions enforced in origin
LG	Legendre Galerkin method
LU	Lower–upper triangular factorization
MEP	Multiparameter eigenvalue problem
O–S	Orr–Sommerfeld equation
S–L	Sturm–Liouville eigenvalue problem
GEP	Generalized eigenvalue problem

QR/QZ	Orthogonal-triangular factorization method for eigenvalues/generalized eigenvalues problems
RGS	Repeated Gram–Schmidt algorithm
rJD	Real variant of Jacobi–Davidson method
TRQI	Tensor Rayleigh quotient iteration
$\mathbf{x}_{int}$	The $N-1$ interior nodes of (CGaussL) partition
Z	The zero matrix of order $N-1$

Chapter 1
General Formulation of Spectral Approximation

Abstract The chapter contains first the general formulation of the spectral approximation as a weighted residual method, i.e., the projection and interpolation operators, test and trial (shape) functions etc. Then, the functional framework of the tau and Galerkin methods based on Chebyshev polynomials is provided, mainly discussing the projection operators. The Chebyshev collocation is introduced in some details. The Chebyshev-Gauss quadrature formulas are reviewed, and the interpolation operator along with the collocation (otherwise called pseudospectral) differentiation matrices are considered. On a regular second order S-L problem, some remarks on the behavior of solutions of Chebyshev collocation, Chebyshev tau and of a Galerkin type method with respect to their order of convergence are conducted.

Keywords Chebyshev polynomials · Galerkin spectral · Phase space differentiation · Tau spectral

> *Spectral methods involve representing the solution to a problem as a truncated series of known functions of the independent variables.*
>
> D. Gottlieb and S.A. Orszag, Numerical Analysis of Spectral Methods: Theory and Applications, SIAM, 1977

1.1 The Tau and Galerkin Spectral Methods

> It's always a numerical nirvana when-for the first time-one observes that adding one or two polynomials to a basis makes the error for smooth problems drops by a factor of 50 even 100.
>
> M. Deville, EPF Laussane, *SIAM News, 37* (2004)

Spectral approximation Let's consider the general linear boundary value problem

C.-I. Gheorghiu, *Spectral Methods for Non-Standard Eigenvalue Problems,*
SpringerBriefs in Mathematics, DOI: 10.1007/978-3-319-06230-3_1,
© The Author(s) 2014

$$\begin{cases} Lu(x) = f(x), x \in (-1, 1), \\ Bu(\pm 1) = 0, \end{cases} \tag{1.1}$$

where L is a linear differential operator and B stands for a set of operators defined on ± 1. We assume that there exists a Hilbert space $(X, (\cdot, \cdot))$ such that $L : X \to X$ can be an unbounded operator and the norm in X is associated to the scalar product. We also assume that $D(L)$ is the definition domain of L and the operators from B make sense to be applied to the functions in $D(L)$. In fact the solution of (1.1) is searched in the subspace $D_B(L) := \{u \in D(L) | Bu = 0\}$. Assume that both $D_B(L)$ and $D(L)$ are dense in X and thus

$$L : D_B(L) \subseteq X \to X. \tag{1.2}$$

With all these we rewrite the problem (1.1) as

$$\begin{cases} u \in D_B(L), \\ Lu = f, \end{cases} \tag{1.3}$$

and assume that it is well formulated, i.e., some criteria such as Lax-Milgram Lemma or inf-sup condition are satisfied (see for instance the monograph [3] or our contribution [18]).

In order to obtain a numerical solution to this problem we first have to approximate the operator L by a family of operators L_N, for some natural number N. Every "discrete" operator L_N will be defined in a finite dimensional subspace X_N of X, which will contain an *approximate solution* of the problem, and its range will be the subspace Z of X. Generally, $X_N \subseteq D_B(L)$ but this is not a necessary condition.

Spectral approximation $u^N \in X_N$ to the solution $u \in X$ will be obtained imposing the vanishing of the projection of the "residual" $L_N u^N - f$ in a finite dimensional subspace Y_N of Z. If Q_N is the orthogonal projection operator $Q_N : Z \to Y_N$, the approximation of order N, u^N will be defined by

$$\begin{cases} u^N \in X_N, \\ Q_N \left(L_N u^N - f \right) = 0. \end{cases} \tag{1.4}$$

For compatibility reasons, the finite dimensional subspaces X_N and Y_N must have the same dimension. They are called the space of *trial* or *shape* functions and respectively *test* functions.

Actually, the *orthogonal projection operator* Q_N will be defined using the scalar product $(\cdot, \cdot)_N$ of the space Y_N, namely

$$\begin{cases} z \in Z, \\ (z - Q_N z, v) = 0, \forall v \in Y_N. \end{cases} \tag{1.5}$$

Thus the spectral approximation (1.4) is equivalent with the *variational approximation*

$$\begin{cases} u^N \in X_N, \\ \left(L_N u^N - f, v\right)_N = 0, \ \forall v \in Y_N. \end{cases} \tag{1.6}$$

In this way the spectral methods can be considered a sub family of the *weighted residual methods*. They can be represented schematically as in the following diagram

$$D_B(L) \subset X \xrightarrow{L} X,$$
$$X_N \subset X \xrightarrow{L_N} Z \subseteq X \xrightarrow{Q_N} Y_N \subset Z.$$

However, we are also interested in some boundary value problems which are particular cases of the very general formulation (1.1), namely *eigenvalue problems*. Thus, we will solve numerically problems of the form

$$\begin{cases} N(\varepsilon, u) = \lambda P(u), \ u := u(x), \ x \in (-1, 1), \ 0 < \varepsilon << 1, \\ U_j^0(u) = \lambda U_j^1(u), \ j = 1, 2, \dots, n. \end{cases} \tag{1.7}$$

Here N and P are ordinary differential expressions of order n and p, respectively, with constant leading coefficients, $n > p$, and $U_j^0(u)$, $U_j^1(u)$ are linear forms containing the variables $u^{(k)}(\pm 1)$ with $k = 0, 1, 2, \dots, n-1$. Problems of this form arise widely in mechanics where usually $n = 4$ and $p = 2$. When the differential expression does not depend on the small parameter ε the general problem (1.7) is thoroughly analyzed in [28]. The authors attach an operator $\mathscr{L} = \mathscr{L}(\lambda)$ to this problem which is defined in $L_2(-1, 1)$ and then the set of values of λ for which $\mathscr{L}(\lambda)$ is not invertible is called the *spectrum* of the problem (1.7). If λ is an eigenvalue, the problem has nontrivial solutions that are called *eigenfunctions* corresponding to λ. Our main aim is to determine accurately some eigenvalues from the spectrum and their corresponding eigenfunctions. Two examples from elasticity and one from hydrodynamics are listed below.

Example 1.1 The analysis of **transverse vibrations of a uniform elastic bar** is based on the differential equation (see for instance the textbook [4])

$$u^{(iv)}(x) = \lambda u(x), \ x \in (0, L), \tag{1.8}$$

where u is the transverse displacement and $\lambda := m\omega^2/EI$; m is the mass per unit length, E is the Young modulus, I is the moment of inertia of the cross section about an axis through centroid perpendicular to the plane of vibration, and ω is the frequency of vibrations. Boundary conditions at each end are usually one of the following type:

- $u = u' = 0$, clamped ends;
- $u = u'' = 0$, simply supported or hinged ends;
- $u'' = u''' = 0$, free ends.

Example 1.2 The analysis of the **buckling of a uniform elastic column** of length L by an axial load P leads to the differential equation (see also [4])

$$-u^{(iv)}(x) = \lambda u''(x), \ x \in (0, L), \tag{1.9}$$

with $\lambda := P/EI$. The physical parameters and the unknown have the same significance as in the above Example. Typical boundary conditions at the ends are clamped or hinged ones.

Example 1.3 A little bit more complicated problem than (1.9) is the following one

$$u^{(iv)}(x) + Ru'''(x) = \lambda u''(x), \ x \in (-1, 1),$$
$$u(\pm 1) = u'(\pm 1) = 0, \tag{1.10}$$

where R is a physical parameter. The eigencondition for this problem is (see [17])

$$\left(R^2 + 4\lambda\right)^{1/2} \left[1 - \frac{\cosh\left(\left(R^2 + 4\lambda\right)^{1/2}\right)}{\cosh(R)}\right] + \frac{2\lambda \sinh\left(\left(R^2 + 4\lambda\right)^{1/2}\right)}{\cosh(R)} = 0. \tag{1.11}$$

When $R = 0$ the problem is self-adjoint and all the eigenvalues are real and less than zero; when R is non-zero the problem is not self-adjoint and the eigenvalues are no longer real, though the real part is negative.

Generalities about *tau* and *Galerkin* spectral methods We briefly introduce the most important properties of Chebyshev polynomials and the ideas behind the tau and the Galerkin spectral methods.

Let $\mathscr{P}_N$ be the set of algebraic polynomials of degree N, for some natural N, and ω the weight function defined on $I := (-1, 1)$, by

$$\omega(x) := \frac{1}{\sqrt{1 - x^2}}. \tag{1.12}$$

It is a continuous, strictly positive and integrable function on I. Let also remember the fundamental space of real Lebesgue square integrable functions defined on the interval I. It is denoted by $L^2_\omega(I)$ and is equipped with the norm $\|v\|_{0,\omega} := \left(\int_{-1}^1 |v(x)|^2 \omega(x)dx\right)^{1/2}$ which is defined by scalar product

$$(u, v)_{0,\omega} := \int_{-1}^1 u(x)v(x)\omega(x)dx. \tag{1.13}$$

The sequence of polynomials defined by

$$T_n(x) := \cos(n \ arc\cos x) = \cosh(n \ arc\cosh x),$$

or, as the unique polynomials satisfying $T_n(\cos\theta) = \cos(n\theta)$ for $n = 0, 1, 2, 3, ...$ will play a key role. They are called *Chebyshev polynomials* and details on them were for the first time systematically provided in the well know monographs [14, 30, 36] (see also [6] for some interesting historical aspects). Thus they are nothing but cosine functions after the change of independent variable $x = \cos\theta$. This property is the root of their widespread importance in the numerical approximation of solutions of non-periodic boundary value problems. Moreover, this change of variables enables a lot of connections with theoretical results concerning Fourier system. A simple computation proves their orthogonality property, i.e., for naturals n and m

$$(T_n, T_m)_{0,\omega} = \frac{\pi}{2} c_n \delta_{nm}. \tag{1.14}$$

Throughout this work δ_{nm} is the Kronecker symbol and

$$c_n := \begin{cases} 2, & n = 0, \\ 1, & n \geq 1, \end{cases} \tag{1.15}$$

By classical Weierstrass approximation Theorem (see for instance [2]) this system is also complete in $L^2_\omega(I)$ and consequently a function u from this space can be expanded as

$$u(x) = \sum_{k=0}^{\infty} \widehat{u}_k T_k(x), \tag{1.16}$$

where

$$\widehat{u}_k = \frac{(u, T_k)_{0,\omega}}{\|T_k\|^2_{0,\omega}} = \frac{2}{c_k}(u, T_k)_{0,\omega}. \tag{1.17}$$

The Chebyshev projection operator is defined as (see [27])

$$P_N u(x) := \sum_{k=0}^{N} \widehat{u}_k T_k(x), \tag{1.18}$$

where $\widehat{u}_k$, $k = \overline{0, N}$ are defined in (1.17). Due to the orthogonality property of Chebyshev polynomials, $P_N u$ represents the orthogonal projection of u on $\mathscr{P}_N$ corresponding to scalar product (1.13). Consequently, we have the important property

$$(P_N u, v)_{0,\omega} = (u, v)_{0,\omega}, \quad \forall v \in \mathscr{P}_N. \tag{1.19}$$

Moreover, due to the completeness property of Chebyshev polynomials we can state that

$$\|u - P_N u\|_{0,\omega} \to_{N \to \infty} 0. \tag{1.20}$$

For the *truncation error* $u - P_N u$ we have the following result (see [8]).

Lemma 1.1 *Let u be a function from the weighted Sobolev space $H^s_\omega(I)$, for a natural s, defined by*

$$H_\omega^s(I) := \left\{ v \in L_\omega^2(I) \,;\; v^{(k)} \in L_\omega^2(I) \,,\; k = \overline{1,s} \right\}. \tag{1.21}$$

Then, we have

$$\|u - P_N u\|_{0,\omega} \leq C\, N^{-s}\, \|u\|_{0,\omega}\,, \tag{1.22}$$

with the constant C independent of N and u.

The proof was first given in [8] for a more general case, i.e., $s \geq 0$. It is also available in [9], Chap. 9 and in [32], Chap. 4. However, details on weighted Sobolev spaces are available in the monographs [1, 9, 20] or even in our contribution [19].

Remark 1.1 Unfortunately, the approximation using Chebyshev projection is optimal only with respect to the scalar product $(\cdot,\cdot)_{0,\omega}$. This is the consequence of a result in [8] which reads

$$\|u - P_N u\|_{1,\omega} \leq C\, N^{2l-s-\frac{1}{2}}\, \|u\|_{0,\omega}\,,\; s \geq l \geq 1, \tag{1.23}$$

where $\|\cdot\|_{p,\omega}$ is the norm in the weighted Sobolev space of order p defined by $\|u\|_{H_\omega^p(I)} := \left(\sum_{k=0}^{p} \|u^{(k)}\|_{L_\omega^2(I)} \right)^{1/2}$. An additional term $2l - \frac{1}{2}$ appears in the power of N. This "non-optimality" led to the introduction of other orthogonal projection with respect to other scalar products ([9], Chaps. 9, 11).

With respect to the functional framework introduced so far, the *Chebyshev tau method* applied to (1.1) is characterized by the following choice

1. $X := L_\omega^2(-1,1)$;
2. $X_N := \{v \in \mathscr{P}_N | \, Bv = 0\}$, and accepts the polynomials T_k, $k = \overline{0,N}$ as a basis in X_N;
3. $Y_N := \mathscr{P}_{N-\beta}$, and accepts the polynomials T_k, $k = \overline{0,N-\beta}$ as a basis in Y_N;

where β stands for the number of boundary conditions involved in B. The projector Q_N is the orthogonal projection operator of X with respect to the scalar product $(\cdot,\cdot)_{0,\omega}$ defined by (1.13). Moreover, $L_N := L$ and Y_N is equipped with the scalar product $(\cdot,\cdot)_N := (\cdot,\cdot)_{0,\omega}$. In order to write down the ChT equations we need first two technical remarks.

Remark 1.2 If the solution is expanded as

$$u^N(x) := \sum_{k=0}^{N} a_k T_k(x), \tag{1.24}$$

its derivatives up to the fourth order can be expressed in the following form (see for instance [17] and the influential paper [31])

$$\left(u^N(x)\right)^{(i)} = \sum_{k=0}^{N} a_k^{(i)} T_k(x), \tag{1.25}$$

where the numerical coefficients $a_k^{(i)}$ are given by

$$c_k a_k^{(1)} = 2 \sum_{p=k+1,\ p+k\ odd}^{N} p a_p,$$

$$c_k a_k^{(2)} = \sum_{p=k+2, p+k even}^{N} p\left(p^2 - k^2\right) a_p,$$

$$c_k a_k^{(3)} = \frac{1}{4} \sum_{p=k+3, p+k\ odd}^{N} p\left[p^2\left(p^2 - 2\right) - 2p^2 k^2 + \left(k^2 - 1\right)^2\right] a_p,$$

$$c_k a_k^{(4)} = \frac{1}{24} \sum_{p=k+4, p+k\ even}^{N} p\left[p^2\left(p^2 - 2^2\right)^2 - 3p^4 k^2 + 3p^2 k^4 - k^2\left(k^2 - 2^2\right)^2\right] a_p.$$

with c_n defined by (1.15).

In this way, the so called *tau differentiation matrices* are defined. Throughout this work the expansion (1.24) will be called *standard Chebyshev basis*.

The condition number of the above fourth order differentiation matrix equals $O(N^8)$ and for large i and j there could be a substantial loss of significance in the bracketed terms (see for instance [9] p. 196 for a detailed discussion).

Remark 1.3 In order to incorporate the boundary conditions in the framework of ChT method, the following equalities are useful

$$T_n^{(k)}(\pm 1) = (\pm 1)^{n+k} \prod_{m=0}^{k-1} \left(n^2 - m^2\right) / 2k - 1)!!, k \geq 1, \tag{1.26}$$

$$N!! := \begin{cases} N(N-2)(N-4)\dots 1, & N\ odd, \\ N(N-2)(N-4)\dots 2, & N\ even. \end{cases}$$

Most of these results follow from the definition of Chebyshev polynomials and trigonometric identities.

Example 1.4 Let's consider the problem (1.1) with $Lu := u^{(iv)} + \lambda^2 u, \lambda \in \mathbb{R}$ supplied with clamped boundary condition, i.e.,

$$\begin{cases} u^{(iv)} + \lambda^2 u = f, \ x \in (-1, 1), \ f \in L^2_\omega(I), \\ u(\pm 1) = u'(\pm 1) = 0. \end{cases} \tag{1.27}$$

If the solution is expanded as in (1.24) the ChT method implies the following linear algebraic system for the unknown coefficients $\{a_k\}_{k=\overline{0,N}}$

$$
\begin{cases}
\int_{-1}^{1} \left[u^{(iv)}(x) + \lambda^2 u(x) \right] T_k(x)\,\omega(x)\,dx = \int_{-1}^{1} f(x)\,T_k(x)\,\omega(x)\,dx, \ k = \overline{0, N-4}, \\
u^N(\pm 1) = \left(u^N \right)'(\pm 1) = 0.
\end{cases}
$$

$$(1.28)$$

In this case $X_N := \left\{ v \in \mathscr{P}_N \mid v(\pm 1) = v'(\pm 1) = 0 \right\}$ and $Y_N := \mathscr{P}_{N-4}$.

Actually, using (1.25) for $i = 4$ and the orthogonality properties of the Chebyshev polynomials, we have to solve the following algebraic system

$$
\begin{cases}
\sum_{k=0}^{N} (\pm 1)^k a_k = \sum_{k=0}^{N} (\pm 1)^{k+1} k^2 a_k = 0, \\
\frac{1}{24 c_k} \sum_{\substack{p=k+4, \ p+k \ even}}^{N} p \left[p^2 \left(p^2 - 2^2 \right)^2 - 3 p^4 k^2 + 3 p^2 k^4 - k^2 \left(k^2 - 2^2 \right)^2 \right] a_p \\
+ \lambda^2 a_k = \widehat{f_k}, \ k = \overline{0, N-4},
\end{cases}
$$

where $\widehat{f_k} := \int_{-1}^{1} f(x) T_k(x) \omega(x) dx, \ k = \overline{0, N}$, are the *Chebyshev coefficients* of f.

In order to solve the above system we can first preallocate a square matrix of order $N + 1$ using `zeros` of MATLAB. Then with two nested `for` loops we can fill in the matrix and the right hand side vector. By Gauss elimination the system is immediately solved. Moreover, the method is *stable*, i.e. $\left\| u^N \right\|_{4,\omega} \leq C \left\| f \right\|_{0,\omega}$ and *convergent* because $\left\| u - u^N \right\|_{4,\omega} \leq C N^{4-m} \left\| u \right\|_{m,\omega}$ with constants C independent of N and u.

For the *classical Galerkin method* we have

1. $X := L_\omega^2 (-1, 1)$;
2. $X_N := Y_N := \{ v \in \mathscr{P}_N \mid Bv = 0 \}$;
3. Find $u^N \in X_N$, such that $\left(L u^N, v \right) = (f, v), v \in X_N$.

As with the tau method, Q_N is the orthogonal projection operator of X with respect to the same scalar product, $L_N := L$ and the scalar product in Y_N, satisfies $(\cdot, \cdot)_N := (\cdot, \cdot)_{0,\omega}$.

Example 1.5 Let's consider that the test and trial bases are built up on the basis

$$
\Phi_k(x) := \begin{cases} T_k(x) - T_0(x), \ k \ even, \\ T_k(x) - T_1(x), \ k \ odd. \end{cases}
$$

$$(1.29)$$

Then the ChG solution to the problem

$$
\begin{cases} -u''(x) = f(x), \ x \in (-1, 1), \\ u(\pm 1) = 0, \end{cases}
$$

reads $u^N(x) := \sum\limits_{k=2}^{N} a_k \Phi_k(x)$ and the ChG formulation becomes

$$-\left(\sum_{k=2}^{N} a_k \Phi_k''(x), \, \Phi_m(x)\right)_{0,\omega} = (f(x), \, \Phi_m(x))_{0,\omega}, \quad m = \overline{2, N}. \qquad (1.30)$$

A simple integration by parts in the left hand side terms leads to the algebraic system

$$A\mathbf{a} = \mathbf{F},$$

where the entries in the matrix A and vector $\mathbf{F}$ are respectively

$$A_{km} := \left(\Phi_k'(x), \, \Phi_m'(x)\right)_{0,\omega}, \quad k, \, m = \overline{2, N},$$

and

$$F_m := (f, \, \Phi_m(x))_{0,\omega}, \quad m = \overline{2, N},$$

and the unknown vector is $\mathbf{a} := (a_2, \ldots, a_N)$. Unfortunately the above basis is not orthogonal and thus the matrix A is fully populated. Additionally, this matrix is not well conditioned and thus one has to take into account possible serious rounding off errors when the system (1.30) is solved. The matrix A and vector $\mathbf{F}$ are otherwise called *stiffness* matrix and *load* vector respectively.

We mention that the most difficult part of Galerkin methods is to build shape and test functions that satisfy the boundary conditions. Moreover, refined variants of Galerkin methods, such as Petrov-Galerkin method, use different test and trial space functions and even operators $L_N \neq L$.

Example 1.6 In [34] the following basis $\{\Phi_k; \, k = \overline{0, N}\}$ where

$$\Phi_k(x) := T_k(x) - T_{k+2}(x), \qquad (1.31)$$

was introduced. It produces the following two scalar products:

$$-\left(\Phi_k, \Phi_j''\right)_{0,\omega} = \frac{\pi}{2} \begin{cases} 4(k+1)(k+2), \, j = k, \\ 8(k+1), \, j = k+2, \, k = 4, \ldots \\ 0, \, otherwise, \end{cases} \qquad (1.32)$$

and

$$(\Phi_k, \Phi_j)_{0,\omega} = \frac{\pi}{2} \begin{cases} c_k + 1, \, j = k, \\ -1, \, j = k \pm 2, \\ 0, \, otherwise, \end{cases} \qquad (1.33)$$

When the test and trial spaces of a Galerkin method will be constructed using this basis, the method will be called of Shen type and denoted ChGS. For $k, j = \overline{0, N}$ these scalar products can be assembled into matrix forms using some basic MATLAB operations such as concatenation, diagonalization of matrices etc. It is clear that the first scalar product above produces an upper triangular matrix and the second one a band matrix. We also observe that these basis functions satisfy the usual Dirichlet boundary conditions $\Phi_k(\pm 1) = 0, k = \overline{0, N}$. An extension of this basis for fourth order problems will be introduced in Chap. 2, Sect. 2.2.

Example 1.7 In [21] another important basis was introduced in order to solve second order problem. It reads $\{\Psi_k; \ k = \overline{0, N}\}$ where

$$\Psi_k(x) := \left(1 - x^2\right) T_k(x). \tag{1.34}$$

It is easy to verify that a Galerkin method based on the following spaces

$$\begin{aligned}
X_N &:= span\left\{\Psi_k| \ \Psi_k(x) := \left(1 - x^2\right) T_k(x), \ k = \overline{0, N}\right\}, \\
Y_N &:= span\left\{\Phi_k| \ \Phi_k(x) := T_k(x) - T_{k+2}(x), \ k = \overline{0, N}\right\},
\end{aligned} \tag{1.35}$$

leads to banded matrices with a conditioning of order $O(N^2)$. They are slightly better conditioned than the matrices produced by (1.31) basis. Such a Galerkin method will be called of the Heinrichs-Shen type and will be denoted by ChGHS. Another extension of this basis for fourth order problems will be introduced in Chap. 2, Sects. 2.1 and 2.2.

1.2 Theoretical Foundation of Spectral Collocation

Chebyshev interpolation is another way to connect the space $L_\omega^2(I)$ with the set of polynomials of a specified degree. Due to exactness reasons the interpolation nodes will be furnished by Chebyshev-Gauss type quadrature formulas, namely

$$\int_{-1}^{1} f(x)\omega(x)\,dx \simeq \sum_{j=0}^{N} f(x_j)\,\omega_j, \tag{1.36}$$

where the choice of the nodes x_j and the weights ω_j lead to different degrees of exactness. The most used formulas are the following (see for instance [9], Chap. 2)

- Chebyshev-Gauss formula, (CGauss)

$$x_j := \cos\frac{(2j + 1)\pi}{2N + 2}, \quad \omega_j := \frac{\pi}{N + 1}, \quad j = \overline{0, N}. \tag{1.37}$$

The nodes x_j are the roots of Chebyshev polynomial T_{N+1} and the formula is exact for polynomials in $\mathscr{P}_{2N+1}$.

- Chebyshev-Gauss-Radau formula, (CGaussR)

$$x_j := \cos \frac{2j\pi}{2N+1}, \quad j = \overline{0, N}, \quad \omega_j := \begin{cases} \frac{\pi}{2N+1}, & j = 0, \\ \frac{\pi}{2N+2}, & j = \overline{1, N}. \end{cases} \quad (1.38)$$

The set of nodes includes 1 and the degree of exactness lowers to $2N$.

- Chebyshev-Gauss-Lobatto formula, (CGaussL)

$$x_j := \cos \frac{j\pi}{N}, \quad j = \overline{0, N}, \quad \omega_j := \begin{cases} \frac{\pi}{2N}, & j = 0, \text{ or } N, \\ \frac{\pi}{N}, & j = \overline{1, N-1}. \end{cases} \quad (1.39)$$

The set of nodes includes both extremities ± 1 and the degree of exactness lowers to $2N - 1$.

With respect to every quadrature formula introduced above, we can define a scalar product and a norm, both *discrete* through the following formulas:

$$(u, v)_N := \sum_{j=0}^{N} \omega_j u\left(x_j\right) v\left(x_j\right), \quad (1.40)$$

$$\|u\|_N := \left(\sum_{j=0}^{N} \omega_j u^2\left(x_j\right)\right)^{1/2}. \quad (1.41)$$

The following result is proved in [32], Chap. 5.

Lemma 1.2

$$\|T_k\|_N = \|T_k\|_{0,\omega}, \quad k = \overline{0, N-1}, \quad (1.42)$$

$$\|T_N\|_N = \begin{cases} \|T_N\|_{0,\omega}, & \text{for } (CGauss), \\ \sqrt{2} \|T_N\|_{0,\omega}, & \text{for } (CGaussL). \end{cases} \quad (1.43)$$

This Lemma is important in showing the uniform equivalence of $\|\cdot\|_N$ and $\|\cdot\|_{0,\omega}$ norms.

Let $I_N u \in \mathscr{P}_N$ be the interpolate of order N with respect to the nodes x_j. In fact we can write

$$I_N u\left(x\right) := \sum_{k=0}^{N} u_k^* T_k\left(x\right), \quad (1.44)$$

where u_k^*, $k = \overline{0, N}$ are the coefficients of discrete Chebyshev series. When we use the (CGaussL) formula we have

$$(I_N u, T_k)_N = \sum_{p=0}^{N} u_p^* (T_p, T_k)_N = \frac{\overline{c}_k \pi}{2} u_k^*, \quad (1.45)$$

where $\bar{c}_k = \begin{cases} 2, & k = 0 \ or \ N, \\ 1, & k = \overline{1, N-1}. \end{cases}$

But $I_N u\left(x_j\right) = u\left(x_j\right)$, $j = \overline{0, N}$, and thus we get

$$(I_N u, T_n)_N = (u, T_n)_N = \sum_{j=0}^{N} \frac{\pi}{\bar{c}_j N} u\left(x_j\right) \cos \frac{nj\pi}{N}. \tag{1.46}$$

The equalities (1.45) and (1.46) provide the formula for the computation of the coefficients of discrete Chebyshev expansion

$$u_k^* = \frac{2}{\bar{c}_k N} \sum_{j=0}^{N} \frac{\pi}{\bar{c}_j} u\left(x_j\right) \cos \frac{kj\pi}{N}. \tag{1.47}$$

The equality (1.47) is the so called *discrete Chebyshev transform* formula. It connects the *physical space* or *the space of values*, in which a solution u is defined by its values on a specified set of nodes, $u(x_i)$, $i = \overline{0, N}$, with the *phase space* or *the space of coefficients* in which the same function u is defined by the coefficients u_k^*, $k = \overline{0, N}$, in the linear combination (1.44). A useful reference for the interpolation using Chebyshev transform and the fast Fourier transform (FFT) is the paper [29] (see also [10]).

For every set of nodes (CGauss), (CGaussR) and (CGaussL) we have (see for instance [9, 32])

Lemma 1.3 *For $u \in H_\omega^m\left(I\right)$, $m \geq 1$ the interpolation error satisfies*

$$\|u - I_N u\|_{0,\omega} \leq C N^{-m} \|u\|_{m,\omega}. \tag{1.48}$$

If $0 \leq l \leq m$, then

$$\|u - I_N u\|_{l,\omega} \leq C N^{2l-m} \|u\|_{m,\omega}, \tag{1.49}$$

and in maximum norm

$$\|u - I_N u\|_{\infty,\omega} \leq C N^{\frac{1}{2}-m} \|u\|_{m,\omega}. \tag{1.50}$$

With the Chebyshev interpolation we can build up the so called *Chebyshev collocation derivative* or sometimes *pseudospectral differentiation (derivative)*. Thus,

$$u'\left(x\right) \simeq (I_N u)'\left(x\right) = \sum_{j=0}^{N} u\left(x_j\right) \Psi_j'\left(x\right) \tag{1.51}$$

where $\Psi_j\left(x\right) \in \mathscr{P}_N$ are the Lagrangian cardinal polynomials associated to the collocation nodes. Thus, on nodes we obtain (see also [20])

$$(I_N u)'\left(x_j\right) = \sum_{i=0}^{N} u\left(x_i\right) \Psi_i'\left(x_j\right), \quad j = \overline{0, N}. \tag{1.52}$$

For the sake of concision we will restrict ourselves to briefly review the ChC method with respect to the nodes (CGaussL) introduced above. The discrete operators of the family $\{L_N\}$ for a natural N, will act on the polynomials from $\mathscr{P}_N$ but generally they will differ from the differential operator L. Actually L_N are obtained from L differentiating in the physical space.

Thus, the *strong collocation method* for (1.1) means to find $u^N \in \mathscr{P}_N$ such that

$$\begin{cases} L_N u^N (x_i) = f(x_i), \ \forall x_i \in (-1, 1), \\ B_N (x_i) = 0, \ \forall x_i \in \{-1, 1\}. \end{cases} \tag{1.53}$$

The nodes x_i belong to (CGaussL) family of nodes and B_N can be identical with B or can be constructed in a similar manner with L_N. Let's denote by $\{\Psi_k \in \mathscr{P}_N, \ k = \overline{0, N}\}$ the Lagrangian basis associated to (CGaussL) nodes, i.e., $\Psi_k (x_j) = \delta_{kj}, j, k = \overline{0, N}$. In this way the *strong collocation solution* is defined as

$$u^N (x) := \sum_{j=0}^{N} u_j \Psi_j (x), \tag{1.54}$$

where $u_j := u^N (x_j), j = \overline{0, N}$, and so the system (1.53) can be rewritten

$$\begin{cases} \sum_{j=0}^{N} u_j L_N \Psi_j (x_i) = f(x_i), \ x_i \neq \{\pm 1\}, \\ \sum_{j=0}^{N} u_j B_N \Psi_j (x_i) = 0, \ x_i = \{\pm 1\}. \end{cases} \tag{1.55}$$

This is a linear system for unknowns $u_j, j = \overline{0, N}$. The entries of its matrix will be computed by differentiation in the physical space. In spite of its simplicity this variant is more difficult to be analyzed theoretically than its weak counterpart.

In order to introduce the *weak collocation method* we will consider the following spaces of trial and test functions

$$X_N := \{v \in \mathscr{P}_N | \ Bv (x_k) = 0, \ \forall x_k \in \{\pm 1\}\}, \tag{1.56}$$

and

$$Y_N := \{v \in \mathscr{P}_N | \ v (x_k) = 0, \ \forall x_k \in \{\pm 1\}\}. \tag{1.57}$$

Thus the space Y_N will be spanned on a subset of $\Psi_k \in \mathscr{P}_N, k = \overline{0, N}$, i.e., of polynomials Ψ_k which vanish on ± 1. The projection operator from (1.4) will be Lagrangian interpolation polynomial defined on the interior (CGaussL) nodes of $(-1, 1)$ and which vanishes on the end nodes ± 1. We also need the "discrete" scalar product defined by

$$(u, v)_N := \sum_{i=0}^{N} u (x_i) v (x_i) \omega_i, \ u, v \in \mathscr{P}_N, \tag{1.58}$$

where $\{x_i,\ i = \overline{0,N}\}$ and $\{\omega_i,\ i = \overline{0,N}\}$ are the nodes and the weights of the associated quadrature formula.

Thus the weak collocation formulation for (1.1) means to find $u^N \in X_N$ such that

$$\left(L_N u^N, \Psi_k\right)_N = (f, \Psi_k),\ \forall \Psi_k,\ k = \overline{1, N-1}, \tag{1.59}$$

or more precisely find $u^N \in X_N$ such that

$$\left(L_N u^N, v\right)_N = (f, v),\ \forall v \in Y_N. \tag{1.60}$$

The error analysis (convergence, stability etc.) for problems of the type (1.60) is available for instance in the monographs [9, 26, 32].

Remark 1.4 Suppose we know the values of a function at several points and we want to approximate its derivative at those points. One way to do this is to find the polynomial that passes through all of the data points, differentiate it analytically, and evaluate this derivative at the grid points. Thus, if $\mathbf{u} := (u(x_0),\ u(x_1),\ ...,\ u(x_n))^T$ is the vector of function values, and $\mathbf{u}' := \left(u'(x_0),\ u'(x_1),\ ...,\ u'(x_n)\right)^T$ is the vector of approximate derivatives obtained by this idea, then there is a matrix D such that

$$\mathbf{u}' = D\mathbf{u}.$$

Such a matrix is called a *collocation differentiation (derivative) matrix*. Approximating differentiation by matrix multiplication is a somewhat expensive computation because a matrix vector multiplication takes $O(n^2)$ operations. Fortunately the matrices used for spectral differentiation have various regularities in them so it is reasonable to hope that they can be exploited to reduce the number of operations needed to accomplish the multiplication. This aspect, the preconditioning of derivative matrices as well as the accuracy of the entries in the derivative matrices were considered in a series of papers such as [5, 13, 15, 16, 35, 38, 40] to quote but a few.

Throughout this paper, as well as throughout our recent papers, we have used the derivative matrices provided in the differentiation matrix suite [39]. The authors pay a lot of attention to the analysis of rounding errors in floating point arithmetic. For instance, in the case of Chebyshev derivative they use the so called "flipping trick" introduced in [12]. In order to avoid floating point cancelation errors in evaluating the difference of abscissas of nodes for large N, they compute only the top half of the matrices and then by symmetry they obtain the lower half.

Remark 1.5 Beyond the above quoted monographs the accuracy of the ChC method was analyzed in [11, 22, 24, 25, 37] to quote but a few.

We end up this section with a simple but suggestive example concerning the observed order of convergence of the classical spectral methods.

Example 1.8 Let's consider the simplest eigenvalue problem, namely

$$\begin{cases} -u'' = \lambda u, \ x \in (0, 1) \,, \\ u\,(0) = u\,(1) = 0. \end{cases} \tag{1.61}$$

It admits the exact eigenpairs $\lambda_n = n^2\pi^2$, and $u_n = B_n \sin(n\pi x)$, $n = 0, 1, 2, 3, \ldots$. If we denote by λ_n^{ChC} the eigenvalues computed by ChC method and introduce a relative error for the first few of them, e.g. five, with

$$e_N := \sum_{i=0}^{5} \left(\lambda_n^{ChC} - \lambda_n \right)^2 \Big/ \sum_{i=0}^{5} \lambda_n^2, \tag{1.62}$$

we can undertake an analysis of the evolution of this error as the order of approximation (cut off parameter) is increased. Using a graphical way to establish the order of a method introduced in [33], we obtain the following dependence

$$\frac{e_{N+1}}{e_N^{0.8}} \simeq 10^{-5.83}, \ for \ N \to 512. \tag{1.63}$$

Actually, we have used the MATLAB code `polyfit` to obtain the regression line in Fig. 1.1. This means that when is applied to the simple problem (1.61) the ChC method is **sublinear**. In fact the relative error increases from highest accuracy of $2.76758545e - 014$ for $N = 2^5$ to $5.19180086e - 011$ when $N = 2^9$. The same analysis performed for ChGS method based on (1.31) basis reveals a slightly better situation, i.e., the observed order equals 0.84. The worse situation is with ChT method. For this one we have found an order which equals 0.65. The analysis is illustrated in Fig. 1.1 with N in the range $2^5 - 2^9$. The conclusion at hand is that the order N of approximation must be chosen with a lot of care (see also in this context Fig. 3.1).

A key aspect in efficient implementation of all variety of spectral methods is the enforcing of the boundary conditions. To this aspect we devote a first remark.

Remark 1.6 Consider the eigenvalue problem (1.7) in one of its simplest form, namely

$$\begin{cases} -u''\,(x) = \lambda u\,(x)\,, \ x \in (-1, 1)\,, \\ \beta u'\,(1) + \alpha u\,(1) = 0, \ \delta u'\,(-1) + \gamma u\,(-1) = 0, \end{cases} \tag{1.64}$$

where the real constants α, β, γ and δ satisfy

$$\begin{aligned} \alpha^2 + \beta^2 \neq 0, \ \alpha\beta \geq 0, \\ \gamma^2 + \delta^2 \neq 0, \ \gamma\delta \leq 0. \end{aligned} \tag{1.65}$$

In [7] are discussed two different ways of treating boundary conditions in collocation methods based on Chebyshev and Legendre polynomials. If the boundary conditions

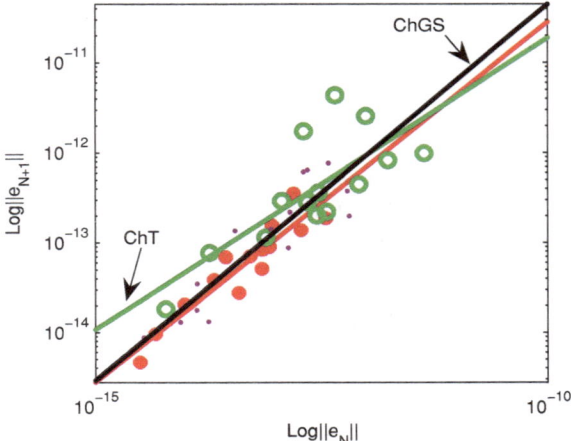

Fig. 1.1 The order of ChC method (*red stars*), ChGS (*magenta dots*) and ChT method (*green circles*) in solving the second order S-L problem (1.61)

in (1.64) are of Dirichlet type, i.e., $\beta = \delta = 0$, the typical spectral collocation scheme consists of collocating the differential equation at the interior nodes and setting to zero the solution on the boundary. This procedure is an **explicit** one and in some situations could be extended to non Dirichlet situation. In an **implicit** procedure the differential equation is collocated at the boundary points also. Thus, the solution is not required to satisfy, and generally it will not satisfy, the boundary conditions exactly. However, it was proved that the boundary conditions can be satisfied up to an error which decays spectrally with the cut off parameter. For high order Neumann boundary conditions, i.e., $u^{(\nu)}(-1) = 0$, $0 \leq \nu \leq l_n$ and $u^{(\nu)}(1) = 0$, $0 \leq \nu \leq r_n$ attached to a problem of order n in [24] is introduced a *weight factor* of the form

$$(1 - x)^{r_n+1} (1 + x)^{l_n+1}, \tag{1.66}$$

along with the general class of Jacobi polynomials. Unfortunately, this technique is inapplicable to some boundary conditions such as mixed or hinged ones where the interpolation is lacunar. We have found the technique conceived in [23] the most flexible and universally applicable one. It is an explicit procedure which is summarized in Sect. 4.3.

References

1. Adams, R.A.: Sobolev Spaces. Academic Press, New York (1975)
2. Atkinson, K., Han, W.: Theoretical Numerical Analysis: A Functional Analysis Framework, 3rd edn. Springer, Berlin (2009)

3. Babuska, I., Aziz, K.: Survey lectures on the mathematical foundation of the finite element method. In: Aziz, A.K. (ed.) The Mathematical Foundations of the Finite Element Method with Application to Partial Differential Equations, pp. 3–359. Academic Press, London (1972)
4. Boyce, W.E., Di Prima, R.C.: Elementary Differential Equations and Boundary Value Problems, 9th edn. Wiley, India (2009)
5. Breuer, K.S., Everson, R.M.: On the errors incurred calculating derivatives using Chebyshev polynomials. J. Comput. Phys. **99**, 56–67 (1992)
6. Butzer, P., Jongmans, F.: P. L. Chebyshev (1821–1894) a guide to his life and work. J. Approx. Theory **96**, 111–138 (1999)
7. Canuto, C.: Boundary conditions in Chebyshev and Legendre methods. SIAM J. Numer. Anal. **23**, 815–831 (1986)
8. Canuto, C., Quarteroni, A.: Approximation results for orthogonal polynomials in Sobolev spaces. Math. Comp. **38**, 67–86 (1982)
9. Canuto, C., Hussaini, M.Y., Quarteroni, A., Zang, T.A.: Spectral Methods in Fluid Dynamics. Springer, New York (1987)
10. Cooley, J.W., Tukey, J.W.: An algorithm for the machine calculation of complex Fourier series. Math. Comput. **19**, 297–301 (1965)
11. Don, W.S., Gottlieb, D.: The Chebyshev-Legendre method: implementing Legendre methods on Chebyshev points. SIAM J. Numer. Anal. **31**, 1519–1534 (1994)
12. Don, W.S., Solomonoff, A.: Accuracy enhancement for higher derivatives using Chebyshev collocation and a mapping technique. SIAM J. Sci. Comput. **18**, 1040–1055 (1997)
13. Elbarbary, E.M.E., El-Sayed, S.M.: Higher order pseudospectral differentiation matrices. Appl. Numer. Math. **55**, 425–438 (2005)
14. Fox, L., Parker, I.B.: Chebyshev Polynomials in Numerical Analysis. Oxford Mathematical Handbooks. O U P, Oxford (1968)
15. Fornberg, B.: A Practical Guide to Pseudospectral Mathods. Cambridge University Press, Cambridge (1998)
16. Funaro, D.: A preconditioning matrix for the Chebyshev differencing operator. SIAM J. Numer. Anal. **24**, 1024–1031 (1987)
17. Gardner, D.R., Trogdon, S.A., Douglass, R.D.: A modified tau spectral method that eliminates spurious eigenvalues. J. Comput. Phys. **80**, 137–167 (1989)
18. Gheorghiu, C.I.: A Constructive Introduction to Finite Elements Method. Quo Vadis, Cluj-Napoca (1999)
19. Gheorghiu, C.I.: Spectral Methods for Differential Problems. Casa Cartii de Stiinta Publishing House, Cluj-Napoca (2007)
20. Gottlieb, D., Orszag, S.A.: Numerical Analysis of Spectral Methods: Theory and Applications. In: SIAM, Philadelphia, Pennsilvania 19103 (1977)
21. Heinrichs, W.: Spectral methods with sparse matrices. Numer. Math. **56**, 25–41 (1989)
22. Heinrichs, W.: Spectral approximation of third-order problems. J. Scientific Comput. **14**, 275–289 (1999)
23. Hoepffner, J.: Implementation of boundary conditions. http://www.lmm.jussieu.fr/~hoepffner/boundarycondition.pdf (2010). Accessed 25 Aug 2012
24. Huang, W., Sloan, D.M.: The pseudospectral method for third-order differential equations. SIAM J. Numer. Anal. **29**, 1626–1647 (1992)
25. Huang, W., Sloan, D.M.: The pseudospectral method for solving differential eigenvalue problems. J. Comput. Phys. **111**, 399–409 (1994)
26. Krasnosel'skii, M.A., Vainikko, G.M., Zabreiko, P.P., Rititskii, YaB, Stetsenko, VYa.: Approximate Solution of Operator Equations. Wolters- Noordhoff, Groningen (1972)
27. Maday, Y.: Analysis of spectral projectors in one-dimensional domains. Math. Comp. **55**, 537–562 (1990)
28. Marletta, M., Shkalikov, A., Tretter, Ch.: Pencils of differential operators containing the eigenvalue parameter in the boundary conditions. Proy. R. Soc. Edinb. **133A**, 893–917 (2003)
29. Monro, D.M.: Interpolation by fast Fourier and Chebyshev transforms. Int. J. Num. Met. Eng. **14**, 1679–1692 (1979)

30. Murty, V.N.: Best approximation with Chebyshev polynomials. SIAM J. Numer. Anal. **8**, 717–721 (1971)
31. Orszag, S.: Accurate solutions of the Orr-Sommerfeld stability equation. J. Fluid Mech. **50**, 689–703 (1971)
32. Quarteroni, A., Valli, A.: Numerical Approximation of Partial Differential Equations. Springer, Berlin (1994)
33. Quarteroni, A., Saleri, F.: Scientific Computing with MATLAB and Octave, 2nd edn. Springer, Berlin (2006)
34. Shen, J.: Efficient spectral-galerkin method II. Direct solvers of second and fourth order equations by using Chebyshev polynomials. SIAM J. Sci. Comput. **16**, 74–87 (1995)
35. Solomonoff, A.: A fast algorithm for spectral differentiation. J. Comput. Phys. **98**, 174–177 (1992)
36. Szegö, G.: Orthogonal Polynomials. American Mathematical Society, New York (1959)
37. Tadmor, E.: The exponential accuracy of Fourier and Chebyshev differencing methods. SIAM J. Numer. Anal. **23**, 1–10 (1986)
38. Trefethen, L.N.: Spectral Methods in MATLAB. In: SIAM, Philadelphia, PA (2000)
39. Weideman, J.A.C., Reddy, S.C.: A MATLAB differentiation matrix suite. ACM Trans. Math. Software **26**, 465–519 (2000)
40. Welfert, B.D.: Generation of pseudospectral differentiation matrices. SIAM J. Numer. Anal. **34**, 1640–1657 (1997)

Chapter 2
Tau and Galerkin Methods for Fourth Order GEPs

Abstract Tau method is detailed mainly for fourth order eigenproblems. For such problems tau differentiation matrices up to fourth order are provided. As some of these problems are self-adjoint a weak (variational) along with a minimization formulation are suggested. The Galerkin method is analyzed with respect to the possibility to choice test and trial functions in order to improve the properties of the differentiation (discretization) matrices, i.e., conditioning, sparsity and symmetry. The non-normality of the differentiation (discretization) matrices is quantified using a scalar measure, i.e., the Henrici's number and the pseudospectrum. The chapter also contains useful hints about the efficient implementation of both methods. A particular attention is paid to the capabilities of tau method to handle GEPs supplied with parameter dependent boundary conditions. The linear stability of some elastic systems as well as the linear hydrodynamic stability of some parallel shear flows (the so called Marangoni-Plateau-Gibbs effect) are analyzed in this context.

Keywords Chebyshev polynomials · Fourth order eigenvalue problems · Galekin spectral · Parameter dependent boundary conditions · Tau spectral

> *In solving a linear eigenvalue problem by a spectral method using $N + 1$ terms in spectral series, the lowest $N/2$ eigenvalues are usually accurate to within a few percent while the larger $N/2$ numerical eigenvalues differ from those of differential equation by such large amounts as to be useless.*
> *Warning 1: the only reliable test is to repeat the calculations with different N and compare the results.*
> *Warning 2: the number of good eigenvalues may be smaller than $N/2$ if the modes have boundary layers, critical levels, or other areas of very rapid change, or when the interval is unbounded.*
> Boyd's EIGENVALUE RULE-OF-THUMB [4]

C.-I. Gheorghiu, *Spectral Methods for Non-Standard Eigenvalue Problems*,
SpringerBriefs in Mathematics, DOI: 10.1007/978-3-319-06230-3_2,
© The Author(s) 2014

2.1 The ChT Method

The best choice of a numerical method is in general problem dependent. Folklore

The standard tau method was introduced in the mid of the last century in the monograph [38] and analyzed in the early stages in the monographs [22, 29], or in the papers [46, 47, 51]. In any of cases, the *standard tau approach*, i.e., the trial (shape) and test functions simply span a certain family of functions (polynomials), has significant disadvantages. First of all, the matrices resulting in the discretization process have an increased condition number, thus computational rounding errors deteriorate the expected theoretical exponential accuracy. Moreover, the discretization matrices are generally fully populated, so efficient algebraic solvers are difficult to apply. Several attempts were made in order to try to circumvent these inconveniences of standard approach. All these attempts are based on the fairly large flexibility in the choice of trial and test functions. In fact, using various weight functions, they are constructed in order to incorporate as much as possible from the boundary data and, at the same time, to reduce the condition number and the bandwidth of matrices. In this respect we mention the papers [5, 14, 20, 64, 65] or our contribution in some joint works [26, 50] (see also [48]). Additionally, the monographs [23, 30] contain details about the spectral tau and Galerkin methods as well as about the collocation (pseudospectral) method. They consider bases of Chebyshev, Hermite and Legendre polynomials as well as of Fourier and *sinc* functions in order to build up the test and trial functions. The well known monograph [4], beyond very subtle observations about the performances and limitations of spectral methods contains an exhaustive bibliography for spectral methods at the level of year 2000. Algorithmic aspects of tau method in solving high order problems, along with some efficient MATLAB codes, have been recently published in [63].

In this section we describe the *tau* variant of spectral methods based on Chebyshev polynomials for some linear fourth order problems. We take into account various boundary conditions.

Crude ChT approximation for transverse vibrations of a uniform elastic bar
Let's consider the Eq. (1.8) introduced in Sect. 1.1 coupled with *mixed boundary conditions*

$$u(0) = u''(0) = 0, \ u(1) = u'(1) = 0, \ L = 1. \tag{2.1}$$

When it is rewritten in $(-1, 1)$ using the change of independent variables $t := 2x - 1$ and is discretized by the above ChT method we get a $N + 1$ order GEP. In this algebraic GEP the differential equation provides $N - 3$ equations, namely

$$16a_k^{(4)} = \lambda, \ k = \overline{0, N - 4},$$

where the entries $a_k^{(4)}$ are defined in Sect. 1.1 with (1.25). The boundary conditions provide the rest of the four equations

$$\sum_{j=0}^{N} (-1)^j a_j = \sum_{j=0}^{N} (-1)^{j+2} \frac{j^2 (j^2 - 1)}{3} a_j = \sum_{j=0}^{N} a_j = \sum_{j=0}^{N} j^2 a_j = 0. \qquad (2.2)$$

In a matrix form these equations can be written as a non Hermitian (singular) GEP

$$A\mathbf{a} = \lambda B\mathbf{a}, \qquad (2.3)$$

where $\mathbf{a} := (a_0 \, a_1 ... a_N)^T$ is the eigenvector. We have solved this GEP by MATLAB code $\texttt{eig}$ and have obtained the following numerical values for the first two eigenvalues $\lambda_1 = 2.377210e + 002$ and $\lambda_2 = 2.496487e + 003$. These values are in perfect agreement with the values provided in [3].

However, due to the homogeneous boundary conditions four diagonal elements from the identity matrix are nullified and thus we get B as a singular matrix. Accordingly four eigenvalues at infinity are produced. Moreover, for N larger than 64 a sort of numerical instability is present and some other spurious (imaginary) eigenvalues may appear. We believe that the non-normality of matrices in the pencil (A,B) is partly responsible for this instability. Fortunately, the MATLAB code $\texttt{eig}$, which implements the $\texttt{qz}$ algorithm explains the presence of the eigenvalues at infinity. In fact the complex pair (α, β), is called an *eigenvalue of the pencil* (A, B) of (2.3) if $\beta A\mathbf{w} = \alpha B\mathbf{w}$, and then $\lambda := \alpha/\beta$, if $\beta \neq 0$. This pair is not unique and if B is *singular*, then

$$(\alpha, \beta) = (1, 0),$$

which corresponds to an infinite eigenvalue.

Remark 2.1 The GEP (1.8)–(2.1) can be cast into the following *variational* and *minimization* problems

$$\textit{find } u \in K \textit{ and real } \lambda \textit{ such that } \int_0^1 u'' v'' dx = \lambda \int_0^1 uv dx, \ \forall v \in K, \qquad (2.4)$$

and

$$\textit{find } \lambda = \min_{u \in K \backslash \{0\}} \frac{\int_0^1 (u'')^2 dx}{\int_0^1 u^2 dx}, \qquad (2.5)$$

where K is the closed subspace of $H^4 (0, 1)$ defined by

$$K = \left\{ v \in H^4 (0, 1) \mid v (0) = v'' (0) = v (1) = v' (1) = 0 \right\}. \qquad (2.6)$$

The minimization formulation assures that the first eigenvalue is a positive one.

We mention that the boundary conditions (2.1) are called *essential* because they are not involved in the variational and minimization formulations. They must be explicitly imposed. On the contrary, the *natural* boundary conditions become explicitly visible in these formulations.

Throughout this work we frequently encounter such singular GEPs. Consequently, we will review two important ideas concerning non-normality of matrix pencils.

Non-normality of spectral approximations One of the main undesirable feature of ChT and ChC methods, appears to be the fact that, due to the non-uniform weight associated with Chebyshev polynomials, they produce by discretization non-symmetric matrices even if the differential problem is self-adjoint.

There are two major concepts with respect to the measure of non-normality of square matrices. The first one is introduced in [35, 36] and provides a *scalar measure* of non-normality (see also the works [10, 19, 21, 28, 40]). For a square complex matrix A the *non-normality ratio* or the *Henrici number* $H(A)$ is defined as

$$H(A) := \left(\varepsilon \left(A^*A - AA^* \right) \right)^{1/2} / \varepsilon(A), \tag{2.7}$$

where A^* is the conjugate transpose of A and $\varepsilon(A)$ stands for the Frobenius norm of A. We have deduced the important estimation

$$0 \le H(A) \le 2^{1/4} \simeq 1.1892, \tag{2.8}$$

with $H(A) = 0$ iff A is normal, i.e., $A^*A - AA^* = 0$.

The second concept, more recently introduced, is that of *pseudospectrum* of a matrix and is systematically treated in [57–60] (see also [15, 16]). Thus for given $N \times N$ matrices A, B and $\varepsilon > 0$ ($\|\cdot\|$ stands for the Euclidean norm) the pseudospectrum or *ε-pseudospectrum* denoted $\Lambda_\varepsilon(A, B)$ is defined by

$$\Lambda_\varepsilon(A, B) := \{\lambda \in \mathbb{C}; \lambda B - A \text{ is singular or } \|\lambda B - A\| < \varepsilon, \ 0 < \varepsilon << 1\}. \tag{2.9}$$

Note that $\Lambda_0(A, B)$ is the usual *spectrum* and the sets $\Lambda_\varepsilon(A, B)$ are nested with respect to ε tending to 0. In [15] it is shown that $\Lambda_\varepsilon(A, B)$ is bounded for nonsingular matrices B and for singular B one has $\Lambda_\varepsilon(A, B)$ equals the whole complex plane for $\varepsilon > \varepsilon^*$ where

$$\varepsilon^* := \min \{\|A\mathbf{u}\| \mid \|\mathbf{u}\| = 1, \ B\mathbf{u} = \mathbf{0}\}.$$

Thus the first conclusion is that the pseudospectra of our singular GEP could cover the entire complex plane. When the entries in our matrices A and B are real, which is the case with our approximations, $\Lambda_\varepsilon(A, B)$ is symmetric with respect to the real axis. This represents an important simplification but to compute the whole pseudospectrum for a singular GEP remains an open problem.

We have used the scalar measure provided by Henrici number and the pseudospectrum as a companion clue in estimating the non-normality of discretizations along the work.

It is also well known that the non-normality is responsible for a high spectral, i.e., with respect to eigenvalues, sensitivity (see for instance [57]).

Fig. 2.1 Pseudospectrum of the fourth order differentiation matrix (1.25). The eigenvalues are the *solid dots* and the boundaries show the Euclidean norm ε-pseudospectrum for $\varepsilon = 10^{-1}$

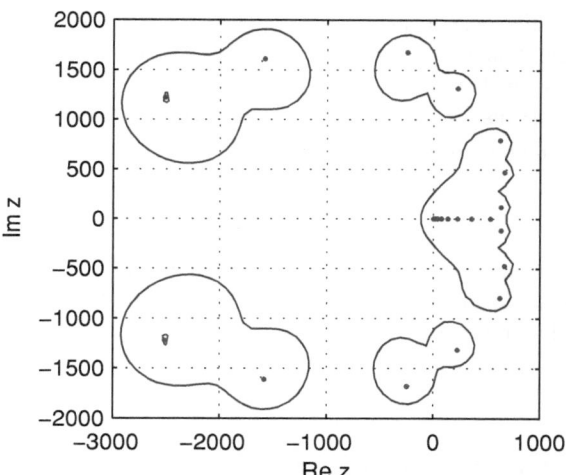

In this context, the numerical experiments reported in our monograph [25] with respect to the *second order differential operator supplied with Dirichlet boundary conditions*, show the following situation.

The discretization matrices of ChT method have a non-normality ratio around 1, and very important, it is quite independent of N, for N in range $2^6 \leq N \leq 2^{10}$. It can be improved to something around 0.8 by suitable choice of trial and test functions.

Roughly, the ChG method produces the most normal discretization matrices. They have a non-normality ratio around 0.1 and this can be lowered to 0.01 for some bases. The conclusion remains valid for orders of approximation considerably increased, i.e., $N = O(2^{10})$. This means matrices fairly close to symmetry. In other words, we can say that the Galerkin discretization tends to best mimic the symmetry properties of the differential operator.

The matrices involved in the ChC method have an intermediate situation. Their non-normality ratio attains 0.4 when $N = 2^8$. All these quantitative estimations of non-normality were confirmed by the pseudospectra of the corresponding matrices.

All in all, one important cause in the lack of accuracy of ChT method seems to be the high non-normality of its finite dimensional counterpart.

For the fourth order differentiation matrix attached to the problem (1.8)–(2.1) we have found a Henrici number which equals 1.0171 and its pseudospectrum (the outmost part) largely extends in the complex plane. This is depicted in Fig. 2.1. Thus the scalar measure along with the pseudospectrum indicate a high departure from normality.

We have to mention that throughout this work by a matrix pseudospectrum we mean the pseudospectrum corresponding to the same matrix scaled by a suitable factor such that the finite part of the spectrum is best visible.

We also have to mention that we will exclusively consider Euclidean norm ε-pseudospectra for $\varepsilon = 10^{-1}, 10^{-2}, \ldots, 10^{-8}$. Thus the corresponding boundaries of the pseudospectra will appear nested from outer to inner. For some matrices the boundaries corresponding to inferior ε coalesce into the dots representing eigenvalues.

Buckling column with hinged boundary conditions Let's consider the Eq. (1.9) introduced in Sect. 1.1 coupled with *hinged boundary conditions*

$$u(0) = u''(0) = 0, \ u(1) = u''(1) = 0, \ L = 1. \tag{2.10}$$

It is again rewritten in $(-1, 1)$ by the same change of independent variables and is discretized by the ChT method. Thus we get a $N + 1$ order singular GEP. In this algebraic GEP the differential equation provides $N - 3$ equations, namely

$$-4a_k^{(4)} = \lambda a_k^{(2)}, \ k = \overline{0, N - 4}, \tag{2.11}$$

where the entries $a_k^{(4)}$ and $a_k^{(2)}$ are defined in Sect. 1.1 with (1.25). The hinged boundary conditions imply the rest of the four equations, namely

$$\sum_{j=0}^{N} (-1)^j a_j = \sum_{j=0}^{N} (-1)^{j+2} \frac{j^2 (j^2 - 1)}{3} a_j = \sum_{j=0}^{N} a_j = \sum_{j=0}^{N} \frac{j^2 (j^2 - 1)}{3} a_j = 0. \tag{2.12}$$

In a matrix form this can be written as a non Hermitian (singular) GEP in a very similar way with (2.3). Solving this GEP with MATLAB built in `eig` the first eigenvalue $\lambda_1 = \pi^2$ is found with five digits precision. We mention that the boundary conditions (2.10) are essential as well.

Buckling column with clamped boundary conditions Consider again the Eq. (1.9) introduced in Sect. 1.1 coupled this time with *clamped boundary conditions*, namely

$$u(0) = u'(0) = 0, \ u(1) = u'(1) = 0, \ L = 1. \tag{2.13}$$

The ChT method involves the same $N - 3$ Eq. (2.11) plus four equations provided by the boundary conditions (2.13). They read

$$\sum_{j=0}^{N} (-1)^j a_j = \sum_{j=0}^{N} (-1)^{j+1} j^2 a_j = \sum_{j=0}^{N} a_j = \sum_{j=0}^{N} j^2 a_j = 0. \tag{2.14}$$

Buckling column with mixed boundary conditions Consider again the Eq. (1.9) coupled this time with *mixed boundary conditions* (2.1). The ChT method casts this problem into an algebraic GEP which concatenates the system (2.11) and the Eq. (2.2).

A slightly modified ChT method In order to improve the conditioning of tau differentiation matrices, in the case of homogeneous clamped boundary conditions, in [34] is proposed the following approach

$$u^N(x) := \sum_{k=0}^{N} a_k \Psi_k(x), \quad \Psi_k(x) := \left(1 - x^2\right)^2 T_k(x), \quad k = \overline{0, N}. \qquad (2.15)$$

This basis a priori satisfies the *homogeneous clamped boundary conditions* $u(\pm 1) = u'(\pm 1) = 0$ and thus it will be also fairly useful for the ChG method. By an elementary, but tedious algebra we can prove the following result.

Lemma 2.1 *For Heinrichs' approximation (2.15) we have*

1. $u^N = \sum\limits_{k=0}^{N+4} b_k T_k$, *where*

$$b_k = \tfrac{c_{k-4}}{16} a_{k-4} + \frac{(c_{k-3}-d_{k-3})-4c_{k-2}}{16} a_{k-2} + \frac{(c_{k-2}-d_{k-2})-4(c_{k-1}-d_{k-1})-6}{16} a_k$$
$$+ \frac{(c_{k-1}-d_{k-1})-4}{16} a_{k+2} + \tfrac{1}{16} a_{k+4};$$
$$(2.16)$$

2. $\left(u^N\right)'' = \sum\limits_{k=0}^{N+2} f_k T_k$, *where*

$$f_k = \frac{c_{k-2}(k+1)(k+2)}{4} a_{k-2} + \frac{4 - k^2 + 3\left(c_{k-1}-d_{k-1}\right)}{2} a_k + \frac{(k-1)(k-2)}{4} a_{k+2};$$
$$(2.17)$$

3. $\left(u^N\right)^{(iv)} = \sum\limits_{k=0}^{N} g_k T_k$, *where*

$$g_k = (k+1)(k+2)(k+3)(k+4) a_k + \frac{5}{c_k} \sum_{\substack{p=k+2,\ p+k\ even}}^{N} p\left(7p^2 - 3k^2 + 20\right) a_p,$$
$$(2.18)$$

with $d_k := \begin{cases} 0, & k < 0 \ or \ k > N, \\ 1, & k = \overline{0, N}, \end{cases}$ *and c_k is defined by (1.15).*

The coefficients with negative indices or indices larger than N were eliminated.

The corresponding matrices have a conditioning of order $O\left(N^4\right)$.

2.2 The ChG Method

In order to formulate the Galerkin methods for fourth order problems some hints are available in [1]. As one of our main aims is to build spectral schemes which rely on well conditioned discretization matrices, we observe that in [53] was introduced a useful expansion, namely $u^N(x) := \sum\limits_{k=0}^{N} u_k \Phi_k(x)$, where

$$\Phi_k(x) := T_k(x) - \frac{2(k+2)}{(k+3)}T_{k+2}(x) + \frac{(k+1)}{(k+3)}T_{k+4}(x), \ k = \overline{0, N}. \qquad (2.19)$$

For this basis we have the following important result.

Lemma 2.2 *The non-zero elements b_{kj}, d_{kj}, and e_{kj}, $k, j = \overline{0, N}$ of the following scalar products are*

1. $\left(\Phi_k, \ \Phi_j\right)_{0,\omega} = \frac{\pi}{2}b_{kj}$, where $b_{kk} = c_k + \frac{4(k+2)^2}{(k+3)^2} + \frac{(k+1)^2}{(k+3)^2}$,

   $b_{kk+2} = b_{k+2k} = -2\left(\frac{k+2}{k+3} + \frac{(k+4)(k+1)}{(k+5)(k+3)}\right)$, $b_{kk+4} = b_{k+4k} = \frac{k+1}{k+3}$;

2. $\left(\Phi_k, \ \Phi_j''\right)_{0,\omega} = \frac{\pi}{2}d_{kj}$, where $d_{kk} = -\frac{8(k+1)(k+2)^2}{k+3}$, $d_{kk+2} = 4(k+1)(k+2)$,

   $d_{kk-2} = 4(k-1)(k+2)$;

3. $\left(\Phi_k, \ \Phi_j^{(iv)}\right)_{0,\omega} = \frac{\pi}{2}e_{kj}$, where

$$e_{kj} = \begin{cases} 16(k+1)^2(k+2)(k+4), \ j = k, \\ \frac{16(k+1)(k+2)}{j+3}\left[k(k+4) + 3(j+2)^2\right], \ j = k+2, k+4, \dots \end{cases} \qquad (2.20)$$

The matrices B, D and E with the entries b_{kj}, d_{kj}, and respectively e_{kj}, $k, j = \overline{0, N}$ are the *Galerkin differentiation matrices* with respect to the Shen basis (2.19). The first one is banded symmetric, the second one is only banded and the third one is upper triangular. The conditioning number of these matrices is of order $O\left(N^4\right)$.

In our approach reported in [50] (see also [48, 49]) a ChGHS method was considered. We have used the Heinrichs' basis $\left\{\Psi_k, \ k = \overline{0, N}\right\}$ as trial basis and the Shen's basis $\left\{\Phi_k, \ k = \overline{0, N}\right\}$ for the test one. We have proved the following important result

Lemma 2.3 *For Φ_i and Ψ_j, $i, j = \overline{0, N}$ defined above and $k = 4, 2, 0$ we have* $\left(\Phi_i, \ \Psi_j^{(k)}\right)_{0,\omega} := \frac{\pi}{2}a_{ij}^k$, *where*

$$a_{ij}^4 = \begin{cases} c_i(i+1)(i+2)(i+3)(i+4), \ j = i, \\ -2i(i+1)(i+2)(i+4), \ j = i+2, \\ i(i+1)^2(i+2), \ j = i+4, \\ 0, \ otherwise, \end{cases} \qquad (2.21)$$

$$a_{ij}^2 = \begin{cases} \frac{c_{i-2}(i+1)(i+2)}{4}, \ j = i-2, \\ \frac{-2i(i+3)+4c_i - 3(c_{i-1}-d_{i-1})}{2}, \ j = i, \\ \frac{3(i+1)(i+2)}{2} + \frac{\delta_{i0}}{2}, \ j = i+2, \\ -(i+1)(i+2), \ j = i+4, \\ \frac{(i+1)(i+2)}{4}, \ j = i+6, \\ 0, \ otherwise, \end{cases} \qquad (2.22)$$

Table 2.1 Condition number for the discretization matrix GD^4

N	$\text{cond}(GD^4)$	$\text{cond}(GD^4)/N^4$	$\text{cond}(PGD^4)$	$\text{cond}(PGD^4)/N^2$
16	3.85×10^3	0.059	2.12×10^1	0.083
64	1.10×10^6	0.065	2.28×10^2	0.070
128	1.89×10^7	0.070	1.11×10^3	0.068
256	3.19×10^8	0.074	4.35×10^3	0.066
512	5.30×10^9	0.077	1.72×10^4	0.066

$$a_{ij}^0 = \begin{cases} \frac{c_{i-4}}{16}, & j = i - 4, \\ -\frac{c_{i-2}(3i+8)-3(c_{i-3}-d_{i-3})}{8(i+3)}, & j = i - 2, \\ \frac{15i+35c_i-22(c_{i-1}-d_{i-1})+5(c_{i-2}-d_{i-2})}{16(i+3)}, & j = i, \\ -\frac{5i+6+4c_i-c_{i-1}+d_{i-1}}{4(i+3)}, & j = i + 2, \\ \frac{15i+22+3c_i}{16(i+3)}, & j = i + 4, \\ -\frac{3i+4}{8(i+3)}, & j = i + 6, \\ \frac{i+1}{16(i+3)}, & j = i + 8, \\ 0, & otherwise, \end{cases} \tag{2.23}$$

with c_i defined in Lemma 2.2 and $d_i := \begin{cases} 0, & if \ i < 0, \\ 1, & if \ i \geq 0. \end{cases}$

Now we want to show the efficiency of our approach. First, if for $k = 0, 2, 4$ we introduce the *Galerkin differentiation matrices* with respect to Heinrichs-Shen bases, namely

$$GD^k := \left(a_{ij}^k\right)_{i,j=\overline{0,N}},$$

with the entries defined respectively by (2.23), (2.22) and (2.21), we observe that GD^4 is upper triangular. Its condition number is of order $O(N^4)$, i.e., similar to the one provided by a finite element discretization. Compared with the result from [53], the condition number is halved. Even the simplest preconditioner, the diagonal one, denoted by P, is effective reducing the condition number to $O(N^2)$. These results are supported by Table 2.1. Similar values are obtained for the banded matrices GD^0 and GD^2.

The accuracy of our approach is illustrated first on a model problem (see also Sect. 3.1).

The celebrated O-S problem reads

$$\begin{cases} \Phi^{(iv)} - 2\alpha^2\Phi'' + \alpha^2\Phi = i\alpha R_e \left[(U - \lambda)\left(\Phi'' - \alpha^2\Phi\right) - U''\Phi\right], & x \in (-1, 1), \\ \Phi(\pm 1) = \Phi'(\pm 1) = 0. \end{cases} \tag{2.24}$$

This GEP occurs in the linear temporal stability analysis of shear flows (see for instance the monographs [17, 18, 52]). Here α denotes the wave number, $i = \sqrt{-1}$,

Fig. 2.2 $\log_{10}|\lambda - \lambda_{ex}|$ when O-S problem is solved by ChGHS method, i.e., based on Shen's basis as test one and Heinrichs' basis as trial one

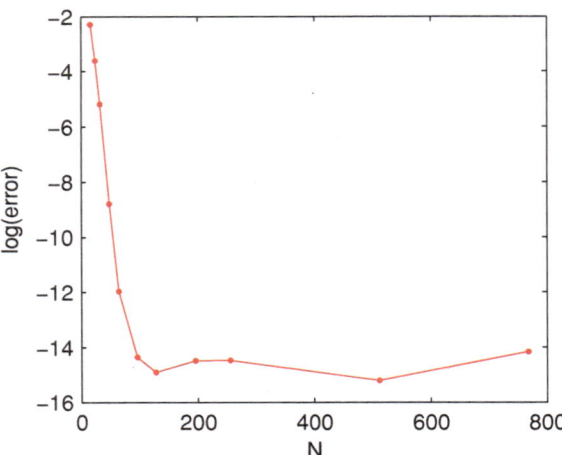

R_e stands for the Reynolds number and the eigenvalues λ are the growth rates. Φ denotes the amplitude of the stream function, while for the basic (unperturbed) profile we take $U(x) := 1 - x^2$, which corresponds to the Poiseuille flow. As pointed out in the seminal paper [45] (see also [14] for computational aspects, or [64, 65]) the ChT method generates spurious eigenvalues due to the way the boundary conditions are imposed. Moreover, numerical stability problems may occur when computations involve large number of polynomials.

Several approaches are proposed in order to alleviate these problems (see [14] and the references therein for spectral methods based on Chebyshev polynomials, or [37, 44] for a Galerkin-Legendre approach). Applying the discretization introduced in Lemma 2.3 we have computed the most unstable mode for $\alpha = 1$ at $R_e = 10^4$. We have to remark that no spurious eigenvalues (at infinity) have been encountered. A discretization of order $N = 64$ achieves an 10^{-12} accuracy. As we can see from Fig. 2.2 this approach is stable. Moreover, increasing N does not lead to a loss of accuracy, computations until $N = O(10^3)$ yielding errors close to the machine precision. We mention that the value furnished in [45] was considered as exact, i.e., $\lambda_{ex} = 0.23752648882 + i0.00373967062$.

Remark 2.2 In order to understand the spectrum behavior of O-S problem (2.24) as the R_e tends to infinity, in [54, 55] it is associated to this problem a lower order problem of the form

$$\begin{cases} -i\varepsilon^2 z'' + U(x)z = \lambda z, \ x \in (-1, 1), \\ z(-1) = z(1) = 0. \end{cases} \tag{2.25}$$

The author assumes that $\varepsilon^2 = (\alpha R)^{-1}$ and thus the problem becomes singularly perturbed. Moreover he shows that this model serves as a fairly suited introduction into the study of the original problem. Actually when $U(x) := (x-\beta)^2$, which means

Fig. 2.3 The spectrum of
problem (2.25) when $N = 64$

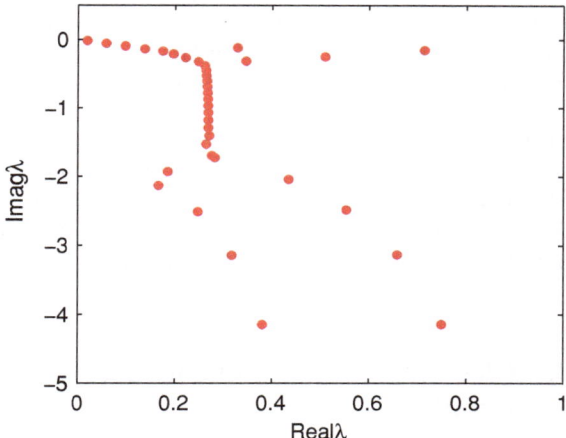

a Couette-Poiseuille profile, the spectrum of problem (2.25) lies in the semi strip

$$\Pi = \{\lambda | Im\lambda < 0, \ 0 < Re\lambda < b\},$$

where $b = (1 - \beta)^2$. Our numerical results for $U(x) := (7x - 1)^2 /64$ and $\varepsilon^2 :=$
10^{-3} are depicted in Fig. 2.3 and plainly confirm this result and the **Y** shape of O-S
spectrum. Similar results but with eigenvalues gathered closely to the real axis were
obtained for ε^2 tending to 10^{-5}. These results have been carried out with a ChG
method using bases of test and trial functions introduced in Sect. 1.1 with (1.29).

The Couette-Poiseuille basic profile is important because in the next Section we
will study the linear hydrodynamic stability of such flows.

Remark 2.3 A Helmholtz boundary value problem analogous to (2.25), i.e.,

$$\begin{cases} -\varepsilon u'' + u = 0, \ x \in (0, 1), \\ u(0) = 0, \ u(1) = 1, \end{cases} \tag{2.26}$$

was considered in [9]. The author found a sort of "maximum principle" in the *phase
space*, in the sense that the Chebyshev coefficients in the representation of the solution
$u^N(x, \varepsilon) := \sum_{k=0}^{N} a_k(\varepsilon) T_k(x)$ are strictly positive for all ε and N. This principle
implies a uniformly bounded solution $u^N(x, \varepsilon)$ in the *physical space (function values
space)*. With this result he shows that for ChC and ChG methods we have

$$\begin{aligned} \|u - u^N\|_{0,\omega} &\le c_1 \min\left\{\tfrac{1}{N^{1/2}}, \ \tfrac{1}{\varepsilon N^{4+1/2}}\right\}, \\ \|u - u^N\|_{L^\infty(-1,1)} &\le c_2 \min\left\{1, \ \tfrac{1}{\varepsilon N^4}\right\}, \end{aligned} \tag{2.27}$$

with the constants c_1 and c_2 independent of ε and N. With respect to ChT method only an estimation of second type holds. However, at least with respect to ChC method we infer that this spectral method performs much better than these estimations predict.

Remark 2.4 It is interesting to observe that in [12, 13] the authors are concerned with Galerkin type methods for fourth order problems based on Jacobi polynomials. They construct bases of such polynomials in order to reduce the condition numbers of discretization matrices to $O\left(N^4\right)$. In this context the importance of the preconditioner introduced above (see Table 2.1) is once more underlined.

2.3 ChT Methods for GEPs with λ Dependent Boundary Conditions

2.3.1 Second Order S-L Problems with Parameter Dependent Boundary Conditions

In [11] the following second order S-L problem is considered

$$\begin{cases} -u''(x) = \lambda u(x),\ 0 < x < \frac{\pi}{2}, \\ u'(0) = \lambda\left(\frac{3}{2}u(0) + u'(0)\right), \\ u'\left(\frac{\pi}{2}\right) = 0. \end{cases} \tag{2.28}$$

Using the crude ChT method formulated in Sect. 1.1 we have obtained the first three eigenvalues 0, $\frac{1}{4}$ and 1 with the machine precision, i.e., the theoretically predicted spectral accuracy was attained. The method casts the differential eigenvalue problem into a singular algebraic one but the computation of the above eigenvalues is not hindered by the infinity eigenvalue.

2.3.2 The Stability of some Elastic Systems

In some buckling problems the eigenvalue parameter appears in the boundary conditions as well as in the differential equation. One such case occurs when one end of the column (bar) is clamped and the other end is free (see for instance [3, 62]). In this case the differential equation (1.9) must be solved subject to the boundary conditions

$$u(0) = u'(0) = u''(1) = u'''(1) + \lambda u'(1) = 0. \tag{2.29}$$

Thus, the crude ChT method using the standard Chebyshev basis, i.e., (1.24) along with (1.25) and (1.26), casts the problem (1.9)–(2.29) into the following algebraic GEP

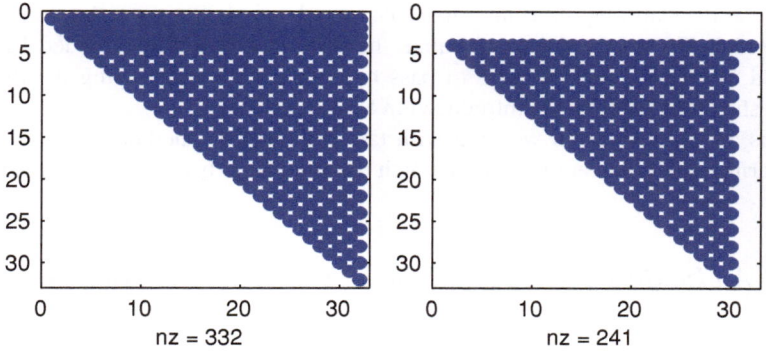

Fig. 2.4 The structure of the matrices A and B in (2.30)

$$\sum_{j=1}^{N+1} (-1)^{j-1} a_j = \sum_{j=1}^{N+1} (-1)^{j+2}(j-1)^2 a_j = \sum_{j=1}^{N+1} \frac{(j-1)^2((j-1)^2-1)}{3} a_j = 0,$$

$$-4\left(\sum_{j=1}^{N+1} \frac{(j-1)^2((j-1)^2-1)((j-1)^2-2^2)}{3\cdot 5} a_j \right) = \lambda \sum_{j=1}^{N+1} (-1)^{j+2}(j-1)^2 a_j,$$

$$\frac{-4}{2\cdot 24} \sum_{p=5,\ p\ odd}^{N+1} p\left[p^2 \left(p^2 - 2^2 \right)^2 \right] a_p = \lambda \sum_{p=3,\ p\ odd}^{N-1} p^3 a_p, \quad (i.e.,\ k=0),$$

$$\frac{-4}{24} \sum_{p=k,\ p+k\ even}^{N+1} p\left[p^2 \left(p^2 - 2^2 \right)^2 - 3p^4 (k-5)^2 \right.$$

$$\left. + 3p^2 (k-5)^2 - (k-5)^2 \left((k-5)^2 - 2^2 \right)^2 \right] a_p$$

$$= \lambda \sum_{p=k-2,\ p+k\ even}^{N-1} p\left(p^2 - (k-5)^2 \right) a_p, \quad k = \overline{5, N+1}$$

$$(2.30)$$

Remark 2.5 In standard notation, one considers approximation of order N and sums, as in (1.24), to have the lower limit $k = 0$ and the upper limit $k = N$. Since MATLAB does not accept a zero index we begin sums with $k = 1$, and consequently our summation will be extended up to $N+1$ term, to get the same order of approximation. With this remark the matrix A is an upper triangular one and B is only quasi upper triangular in (2.30). Their structures are available in Fig. 2.4.

We have obtained as the first eigenvalue λ_1 the numerical value 2.467401 which is a six digits correct approximation of the exact value $(\pi/2)^2$. The next two computed eigenvalues are 2.220660e+001 and 6.168502e+001.

Remark 2.6 From physical point of view it means that for $\lambda < \lambda_1$ the rectilinear form of the bar is a *stable* equilibrium state, i.e., this form is stable with respect to the "small" arbitrary perturbations. If $\lambda > \lambda_1$ this equilibrium state is *unstable*. The non null eigensolution corresponding to λ_1 defines the character of the deformation as soon as the stable equilibrium is lost (see for a detailed discussion the well known monograph [39]). In order to visualize this first eigensolution provided by ChT method in phase space we have to pass to the physical space using the discrete Chebyshev transform (1.47) introduced in Sect. 1.2.

This additional step in working with tau and Galerkin methods is a drawback comparing with collocation method which operates directly in the physical space.

2.3.3 A Modified ChT Method for a Particular O-S Problem

Different numerical methods are usually more similar than they are distinct. Brenner and Scott, The Mathematical Theory of F E M, 2008

The hydrodynamic stability problem analyzed below describes the appearance of long wave instability in a thin film of isothermal liquid induced by a surface tension gradient. The liquid film flows over a rigid plane inclined at the angle β with respect to the horizontal. The film is bounded above by a gas (the atmosphere) which exerts a shear stress on the interface. However, this stress is negligible when compared with the stress due to the surfactant and consequently will not be taken into account. We introduce the scales for velocity, pressure, length and time and examine the *temporal* stability of flow using the standard linear stability analysis (see for an analogous discussion [2, 56]). From the mathematical point of view we have to solve the GEP

$$L\Phi = 0, \ x \in (0, 1), \tag{2.31}$$

$$\Phi(1) = \Phi'(1) = 0, \tag{2.32}$$

$$\Phi''(0) + \alpha^2 \Phi(0) + NU''(0) = 0, \tag{2.33}$$

$$\Phi'''(0) + i\alpha [R_e (\lambda - U(0)) + 3\alpha i] \Phi'(0)$$
$$+ i\alpha [2\cot(\beta) + \alpha^2 Ca + (\lambda - U(0)) R_e U'(0)] N = 0, \tag{2.34}$$

where

$$L(\cdot) := \left((D^2 - \alpha^2)^2 - i\alpha R_e (U(D^2 - \alpha^2) - U'') \right) (\cdot) + i\alpha R_e \lambda (D^2 - \alpha^2) (\cdot), \tag{2.35}$$

and $N := \frac{\Phi(0)}{\lambda - U(0)}$. Here $U(x) = (1 - x^2) + (1 - x)\tau$ is the velocity profile of the unperturbed basic flow, R_e stands again for the Reynolds number, α denotes the wave number of the disturbance, τ stands for the dimensionless surface stress and $-i\alpha R_e \lambda$ determines the exponential time dependence via $\varphi(x, z, t) := \Phi(x) e^{i\alpha(z - \lambda t)}$. $\overline{\lambda}_r = Re(\lambda)$ provides the phase speed and $\alpha \overline{\lambda}_i = \alpha \text{Im}(\overline{\lambda})$ gives the growth rate of instability. Here $\bar{\lambda}$ stands for the most unstable mode, i.e., the eigenvalue having the largest imaginary part.

If the Reynolds number R_e is less than the "critical value", the growth rate of instability is negative and thus the flow is *linearly stable*. Using long wave asymptotic expansion we have obtained for the critical Reynolds number the following estimate

$$R_{e,cr} = \frac{5}{2} \cot{(\beta)} \frac{1}{2 + \tau}. \tag{2.36}$$

For the phase speed we have got

$$\bar{\lambda}_r = 2 + \tau. \tag{2.37}$$

The differential operator $L\,(\cdot)$ introduced above is in fact identical to the one in the well known O-S equation (2.24). The differences appear at the last two boundary conditions where the spectral parameter λ occurs nonlinearly. Fortunately this non-linearity has a special type and can easily be removed. Concretely, because our basic profile is parabolic, $U''\,(0) \neq 0$, we can eliminate N from (2.34) and obtain

$$U''\,(0)\,\Phi'''\,(0) + i\alpha\,[R_e\,(\lambda - U\,(0)) + 3\alpha i]\,\Phi'\,(0)$$
$$-i\alpha\,[2 \cot{(\beta)} + \alpha^2 Ca + (\lambda - U\,(0))\,R_e U'\,(0)]\,[\Phi''\,(0) + \alpha^2 \Phi\,(0)] = 0. \tag{2.38}$$

The Eq. (2.33) can be rewritten as

$$(\lambda - U(0))\left[\Phi''\,(0) + \alpha^2 \Phi(0)\right] + \Phi(0)U''\,(0) = 0. \tag{2.39}$$

Thus the initial problem has been transformed into a new one with boundary conditions depending only linearly on λ. Consequently in the subsequent analysis we will consider only the modified boundary conditions (2.38) and (2.39).

As mentioned above an accurate numerical method to compute the eigenvalues of this new problem is needed. For this reason we have considered a Chebyshev spectral one in our previous paper [27]. Although the above modifications simplify our problem, the implementation of boundary conditions is still a challenging task. It is not elementary at all to implement them in a collocating or Galerkin formulation, while in the tau method this can be done more easily.

In the tau approach, the space on which the residual is projected has a lower dimension and the boundary conditions are imposed explicitly. For the classical ChT discretization of the problem (2.31), (2.32), (2.38), (2.39) we first transform it linearly on $(-1, 1)$ and then make use of the following spaces and bases

$$X_N := \mathscr{P}_{N+4} = span\left\{T_k,\ k = \overline{0, N+4}\right\},$$
$$Y_N := \mathscr{P}_N = span\left\{T_k,\ k = \overline{0, N}\right\}. \tag{2.40}$$

The projection mentioned above is orthogonal with respect to the $L^2_\omega(-1, 1)$ scalar product and thus the ChT discretization for the problem at hand reads

$$\begin{cases} \Phi_N \in X_N, \\ (L_t\Phi_N, \varphi)_{0,\omega} = 0, \ \forall\varphi \in Y_N, \\ B.C.'s, \end{cases} \tag{2.41}$$

where L_t denotes the transformed differential operator on $(-1, 1)$ and by $B.C.$ we mean an explicit impose of boundary conditions in ± 1. In fact we can split L_t as $L_t(\cdot) := L_l(\cdot) + \lambda L_r(\cdot)$, where

$$\begin{aligned} L_l(\cdot) &:= \left((2^2D^2 - \alpha^2)^2 - i\alpha R_e\left(U\left(2^2D^2 - \alpha^2\right) - U''\right)\right)(\cdot), \\ L_r(\cdot) &:= i\alpha R_e\left(2^2D^2 - \alpha^2\right)(\cdot). \end{aligned} \tag{2.42}$$

Taking $\Phi_N := \sum_{j=0}^{N+4} a_j T_j$ the equality $(L_t\Phi_N, \varphi)_{0,\omega} = 0$, $\forall\varphi \in Y_N$, implies $N + 1$ equations for the coefficients a_j, $j = \overline{0, N+4}$ and the spectral parameter λ. The remaining four equations are given by the boundary conditions in ± 1. Hence we have obtained a singular GEP,

$$A\mathbf{a} = \lambda B\mathbf{a}, \tag{2.43}$$

with the vector $\mathbf{a}$ containing the coefficients a_j, $j = \overline{0, N+4}$. The matrices A and B correspond to the discretization of operators L_l and respectively $-L_r$ by ChT method, i.e., by formulas of type (1.25). We observe that

$$rank(A) = rank(B) + 2,$$

i.e., $\beta = 2$, because the contribution of boundary conditions (2.32), is visible only in A. Unfortunately the discretization matrices determined by this method are quite badly conditioned and not sparse. Even though this method has *theoretically* a spectral accuracy, it is strongly affected by rounding off errors. Moreover, whenever a QZ type methods is used to solve (2.43) it generates two spurious eigenvalues (at infinity).

Consequently, we have improved this ChT method as follows. We introduce the space of *trial (shape) functions*

$$\widetilde{X}_N := span\left\{\Theta_i, \ i = \overline{0, N+2}\right\}, \tag{2.44}$$

where every function

$$\Theta_i(x) := (1 - x)^2 T_i(x), \ i \geq 0, \tag{2.45}$$

a priori satisfies the *essential* boundary conditions in $x = 1$. Therefore we are looking for a solution of the form

$$\Phi_N := \sum_{j=0}^{N+2} a_j\Theta_j, \tag{2.46}$$

and it is clear that $\widetilde{X}_N = \{v \in \mathscr{P}_{N+4}; \; v(1) = v'(1) = 0\}$.

For the definition of *test functions* we use the functions below. First, we define

$$\Psi_i^1(x) := \frac{2}{\pi} d_i^N \left[\frac{2i+3}{4(i+1)} T_i(x) - T_{i+1}(x) + \frac{2i+1}{4(i+1)} T_{i+2}(x) \right], \; i \geq 0, \quad (2.47)$$

where $d_i^N := 1$ if $0 \leq i \leq N$ and $d_i^N := 0$ otherwise. Now let's introduce the set of functions

$$\Psi_i^{k+1}(x) := \frac{1}{2i+k+2} \left\{ \Psi_i^k(x) + \Psi_{i+1}^k(x) \right\}, \; i \geq 0 \; and \; k \geq 1. \quad (2.48)$$

The choice of the test functions is justified by the following

Lemma 2.4 *For every $k \geq 1$ we have*

$$\widetilde{Y}_N := span \left\{ \Psi_i^k, \; i = \overline{0, N} \right\} = \{v \in \mathscr{P}_{N+2}; \; v(1) = v'(1) = 0\}. \quad (2.49)$$

Proof The case $k = 1$ is obvious. For $k > 1$ the mathematical induction can be applied directly.

Consequently, the *modified ChT formulation* of (2.41) reads

$$\begin{cases} \Phi_N \in \widetilde{X}_N, \\ (L_t \Phi_N, \varphi)_{0,\omega} = 0, \; \forall \varphi \in \widetilde{Y}_N, \\ B.C.'s, \end{cases} \quad (2.50)$$

but now the $B.C.'s$ stands only for the boundary conditions in -1 , which still have to be imposed explicitly.

For the discretization of L_t we take Ψ_i^5, $i = \overline{0, N}$ as test functions and construct the following discretization matrices. This choice is motivated by the following result

Lemma 2.5 *Let Θ_j and Ψ_i^5 be as above. For all $i = \overline{0, N}$ and $j \geq 0$ we have* $\left(\Psi_i^5, \Theta_j^{(iv)} \right) = d_j^{N+2} a_{ij}$, *where*

$$a_{ij} = \begin{cases} \frac{i+4}{4(2i+5)}, & j = i+2, \\ -\frac{(i+2)(i+4)}{(2i+5)(2i+7)}, & j = i+3, \\ \frac{i+2}{4(2i+7)}, & j = i+4, \\ 0, & otherwise, \end{cases} \quad (2.51)$$

and $d_j^N := 1$ if $0 \leq i \leq N$ and $d_j^N := 0$ otherwise.

Proof The proof is quite elementary but tedious and is based on the orthogonal properties of Chebyshev polynomials.

Table 2.2 Most unstable mode, crude (classical) ChT approach ($\tau = 1.75$)

N	$\text{Re}(\overline{\lambda})$	$\text{Im}(\overline{\lambda})$	$\log_{10}|\overline{\lambda} - \lambda_{ex}|$
8	3.7499999800	$-0.446\text{E}-10$	-7.70
64	3.7499999805	$-0.553\text{E}-05$	-5.26
96	3.7499999854	$-0.143\text{E}-03$	-3.64
128	3.7500041460	$-0.508\text{E}-02$	-2.29

Table 2.3 Most unstable mode, modified ChT approach ($\tau = 1.75$).

N	$\text{Re}(\overline{\lambda})$	$\text{Im}(\overline{\lambda})$	$\log_{10}|\overline{\lambda} - \lambda_{ex}|$
8	3.7499999800	$-0.261\text{E}-11$	-7.70
64	3.7499999805	$-0.828\text{E}-10$	-7.71
96	3.7499999854	$-0.844\text{E}-08$	-8.02
128	3.7500041460	$0.970\text{E}-06$	-5.37

Remark 2.7 In this approach, the discretization matrix for the fourth order derivative is banded and it is better conditioned than the corresponding matrix (1.25) which is only upper triangular. Consequently, the method featured more stability. However, similar discretization matrices are obtained for lower order derivatives or other operators.

Remark 2.8 Fairly similar ideas can be used in order to find optimal bases in test and trial spaces corresponding to other homogeneous boundary conditions.

The efficiency of our approach results comparing Tables 2.2 and 2.3.

The exact value of the most unstable mode was $\lambda_{ex} = 3.75$. Moreover, in [31, 32] our numerical values for the most unstable mode of (2.31)–(2.34) have been confirmed.

Remark 2.9 The GEP (2.43) is a representative singular one. The occurrence of spurious eigenvalues in such problems was discussed by many authors and the reader is referred to the papers [24, 41, 43, 65] and the references therein. Unfortunately the proposed remedies are some ad-hoc ones. Thus, in [24] was introduced a method that splits the matrices of the pencil by eliminating β coefficients a_k ("degrees of freedom") from the left hand side matrix using β boundary conditions rows. Unfortunately, it is not clear at all which coefficients a_k can be eliminated such that the region of interest in the spectrum remains unaffected. Also in the monograph [52] is used a trick to avoid eigenvalues at infinity. The zero rows in the right hand side matrix are perturbed introducing large complex entries. Instead, we suggest some *JD* type methods which act on a specified target from the whole spectrum and systematically eliminate spurious eigenvalues. However, the issue is still a research topic (see for instance the recent work [33]).

Remark 2.10 A novel O-S like problem for gravity-driven turbulent open-channel flows over a granular erodible bed has been recently considered in [7]. The authors encounter the same difficulties we have noticed above and in the framework of a Galerkin scheme they propose an attractive technique to introduce second order boundary conditions (see also [6, 8]). Instead, we have explicitly enforced the second and third order boundary conditions in $x = -1$.

Remark 2.11   Within a sophisticated functional framework in [61, 62] and latter in [42] the authors consider pencils of differential operators containing the eigenvalue parameter in the boundary conditions. In these papers the authors take our problem (2.31)–(2.34) as a remarkable example. They observe that the differential operator L_r along with boundary conditions (2.32) is invertible. Making use of its inverse, they find out a linearization of our problem, i.e., an eigenvalue problem which does not contain the spectral parameter in the boundary conditions. Moreover, they show that the spectrum and the set of eigenfunctions of the original problem and of its linearization coincide. Unfortunately, this linearization is not useful from numerical point of view because the degree of differentiation is increased by one unit and the linearized boundary conditions contain some inconvenient integral operators.

References

1. Bjoerstad, P.E., Tjoestheim, B.P.: Efficient algorithms for solving a fourth-equation with the spectral-Galerkin method. SIAM J. Sci. Stat. Comput. **18**, 621–632 (1997)
2. Blyth, M.G., Pozrikidis, C.: Effect of surfactant on the stability of film flow down an inclined plane. J. Fluid Mech. **521**, 241–250 (2004)
3. Boyce, W.E., Di Prima, R.C.: Elementary Differential Equations and Boundary Value Problems, 9th edn. Wiley, India (2009)
4. Boyd, J.P.: Chebyshev and Fourier Spectral Methods, 2nd edn. DOVER Publications, Inc., New York (2000)
5. Cabos, Ch.: A preconditioning of the tau operator for ordinary differential equations. ZAMM **74**, 521–532 (1994)
6. Camporeale, C., Ridolfi, L.: Ice ripple formation at large Reynolds numbers. J. Fluid Mech. **694**, 225–251 (2012)
7. Camporeale, C., Canuto, C., Ridolfi, L.: A spectral approach for the stability analysis of turbulent open channel flows over granular beds. Theor. Comput. Fluid Dyn. **26**, 51–80 (2012)
8. Camporeale, C., Mantelli, E., Manes, C.: Interplay among unstable modes in films over permeable wals. J. Fluid Mech. **719**, 527–550 (2013)
9. Canuto, C.: Spectral Methods and a Maximum Principle. Math. Comput. **51**, 615–629 (1988)
10. Chaitin-Chatelin, F., Frayssé, V.: Lectures on Finite Precision Computation. SIAM, Philadelphia (1996)
11. Chanane, B.: Computation of the eigenvalues of Sturm-Liouville problems with parameter ependent boundary conditions using the regularized sampling method. Math. Comput. **74**, 1793–1801 (2005)
12. Doha, E.H.: On the coefficients of differential expansions and derivatives of Jacobi polynomials. J. Phys. A: Math. Gen. **35**, 3467–3478 (2002)
13. Doha, E.H., Bhrawy, A.H.: Efficient spectral-Galerkin algorithms for direct solution of fourth-order differential equations using Jacobi polynomials. Appl. Numer. Math. **58**, 1224–1244 (2008)
14. Dongarra, J.J., Straughan, B., Walker, D.W.: Chebyshev tau- QZ algorithm for calculating spectra of hydrodynamic stability problems. Appl. Numer. Math. **22**, 399–434 (1996)
15. van Dorsslaer, J.L.M.: Pseudospectra for matrix pencils and stability of equilibria. BIT Numer. Math. **37**, 833–845 (1997)

16. van Dorsslaer, J.L.M.: Several concepts to investigate strongly nonnormal eigenvalue problems. SIAM J. Sci. Comput. **24**, 1031–1053 (2003)
17. Drazin, P.G., Reid, W.H.: Hydrodynamic Stability. Cambridge University Press, Cambridge (1981)
18. Drazin, P.G., Beaumont, D.N., Coaker, S.A.: On Rossby waves modified by basic shear, and barotropic instability. J. Fluid Mech. **124**, 439–456 (1982)
19. Eberlein, P.J.: On measures of non-normality for matrices. Amer. Math. Monthly **72**, 995–996 (1965)
20. El-Daou, M.K., Ortiz, E.L., Samara, H.: A unified approach to the tau method and Chebyshev series expansion techniques. Comput. Math. Appl. **25**, 73–82 (1993)
21. Elsner, L., Paardekooper, M.H.C.: On measure of nonnormality of matrices. Linear Algebra Appl. **92**, 107–124 (1987)
22. Fox, L., Parker, I.B.: Chebyshev Polynomials in Numerical Analysis. Oxford Mathematical Handbooks. Oxford University Press, Oxford (1968)
23. Funaro, D.: Polynomial Approximation of Differential Equations. Springer, Berlin Heidelberg (1992)
24. Gardner, D.R., Trogdon, S.A., Douglass, R.D.: A modified tau spectral method that eliminates spurious eigenvalues. J. Comput. Phys. **80**, 137–167 (1989)
25. Gheorghiu, C.I.: Spectral Methods for Differential Problems. Casa Cartii de Stiinta Publishing House, Cluj-Napoca (2007)
26. Gheorghiu, C.I., Pop, S.I.: On the Chebyshev-tau approximation for some singularly perturbed two-point boundary value problems. Rev. Roum. Anal. Numer. Theor. Approx. **24**, 117–124 (1995)
27. Gheorghiu, C.I., Pop, S.I.: A Modified Chebyshev-tau method for a hydrodynamic stability problem. In: Proceedings of ICAOR, Cluj-Napoca, vol. II, pp. 119–126 (1997)
28. Golub, G.H., van der Vorst, H.A.: Eigenvalue computation in the 20th century. J. Comput. Appl. Math. **123**, 35–65 (2000)
29. Gottlieb, D., Orszag, S.A.: Numerical Analysis of Spectral Methods: Theory and Applications, p. 19103. SIAM, Philadelphia, Pennsilvania (1977)
30. Gottlieb, D., Hussaini, M.Y., Orszag, S.A.: Theory and applications of spectral methods. In: Voigt, R.G., Gottlieb, D., Hussaini, M.Y. (eds.) Spectral Methods for Partial Differential Equations, pp. 1–54. SIAM-CBMS (1984).
31. Greenberg, L., Marletta, M.: Numerical methods for higher order Sturm-Liouville problems. J. Comput. Appl. Math. **125**, 367–383 (2000)
32. Greenberg, L., Marletta, M.: Numerical solution of non-self-adjoint Sturm-Liouville problems and related systems. SIAM J. Numer. Anal. **38**, 1800–1845 (2001)
33. Hagan, J., Priede, J.: Capacitance matrix technique for avoiding spurious eigenmodes in the solution of hydrdynamic stability problems by Chebyshev collocation method. arXiv:1207.0388v2[physics.com-php]. Accessed 14 Dec 2012
34. Heinrichs, W.: A stabilized treatment of the biharmonic operator with spectral methods. SIAM J. Sci. Stat. Comput. **12**, 1162–1172 (1991)
35. Henrici, P.: Bounds for iterates, inverses, spectral variation and fields of values of non-normal matrices. Numer. Math. **4**, 24–40 (1962)
36. Henrici, P.: Discrete Variable Methods in Ordinary Differential Equations. Wiley, New York, London (1962)
37. Kirkner, N.P.: Computational aspects of the spectral Galerkin FEM for the Orr-Sommerfeld equation. Int. J. Numer. Meth. Fluids **32**, 119–137 (2000)
38. Lanczos, C.: Applied Analysis. Prentice Hall Inc., Englewood Cliffs (1956)
39. Landau, L., Lifchitz, E.: Théorie de L'Élasticité. Édition Mir, Moscou (1967)
40. Lee, S.L.: A practical upper bound for departure from normality. SIAM J. Matrix Anal. Appl. **16**, 462–468 (1995)
41. Lindsay, K.A., Odgen, R.R.: A practical implementation of spectral methods resistant to the generation of spurious eigenvalues. Intl. J. Numer. Fluids **15**, 1277–1294 (1992)

42. Marletta, M., Shkalikov, A., Tretter, Ch.: Pencils of differential operators containing the eigenvalue parameter in the boundary conditions. Proc. R. Soc. Edinb. **133A**, 893–917 (2003) NULL
43. McFaden, G.B., Murray, B.T., Boisvert, R.F.: Elimination of spurious eigenvalues in the Chebyshev tau spectral methods. J. Comput. Phys. **91**, 228–239 (1990)
44. Melenk, J.M., Kirchner, N.P., Schwab, C.: Spectral Galerkin discretization for hydrodynamic stability problems. Computing **65**, 97–118 (2000)
45. Orszag, S.: Accurate solutions of the Orr-Sommerfeld stability equation. J. Fluid Mech. **50**, 689–703 (1971)
46. Ortiz, E.L.: The tau method. SIAM J. Numer. Anal. **6**, 480–492 (1969)
47. Ortiz, E.L., Samara, H.: An operational approach to the tau method for the numerical solution of non-linear differential equations. Computing **27**, 15–25 (1981)
48. Pop, I.S.: A stabilized approach for the Chebyshev-tau method. Stud. Univ. Babes-Bolyai, Math. **42**, 67–79 (1997)
49. Pop, I.S.: A stabilized Chebyshev-Galerkin approach for the biharmonic operator. Bul. Stint. Univ. Baia-Mare Ser. B **14**, 335–344 (2000)
50. Pop, I.S., Gheorghiu, C.I.: A Chebyshev-Galerkin method for fourth order problems. In: Proceedings of ICAOR, Cluj-Napoca, vol. II, pp. 217–220 (1997)
51. Roos, H.G., Pfeiffer, E.: A convergence result for the tau method. Computing **42**, 81–84 (1989)
52. Schmid, P.J., Henningson, D.S.: Stability and Transition in Shear Flows. Springer, New York (2001)
53. Shen, J.: Efficient spectral-Galerkin method II. Direct solvers of second and fourth order equations by using Chebyshev polynomials. SIAM J. Sci. Comput. **16**, 74–87 (1995)
54. Shkalikov, A.A.: Spectral portrait of the Orr-Sommerfeld operator with large Reynolds numbers. arXiv:math-ph/0304030v1 (2003). Accessed 25 Aug 2010
55. Shkalikov, A.: Spectral portrait and the resolvent growth of a model problem associated with the Orr-Sommerfeld equation. arXiv:math.FA/0306342v1 (2003). Accessed 25 Aug 2010
56. Smith, M.K.: The mechanism for long-wave instability in thin liquid films. J. Fluid Mech. **217**, 469–485 (1990)
57. Trefethen, L.N.: Pseudospectra of linear operators. SIAM Rev. **39**, 383–406 (1997)
58. Trefethen, L.N.: Computation of pseudospectra. Acta Numerica **9**, 247–295 (1999)
59. Trefethen, L.N., Trummer, M.R.: An instability phenomenon in spectral methods. SIAM J. Numer. Anal. **24**, 1008–1023 (1987)
60. Trefethen, L.N., Embree, M.: Spectra and Pseudospectra. The Behavior of Nonnormal Matrices. Princeton University Press, Princeton and Oxford (2005)
61. Tretter, Ch.: A linearization for a class of λ-nonlinear boundary eigenvalue problems. J. Math. Anal. Appl. **247**, 331–355 (2000)
62. Tretter, Ch.: Boundary eigenvalue problems for differential equations $N\eta = \lambda P\eta$ with λ-polynomial boundary conditions. Integr. J. Diff. Equat. **170**, 408–471 (2001)
63. Trif, D.: Operatorial tau method for higher order differential problems. Br. J. Math. Comput. Sci. **3**, 772–793 (2013)
64. Zebib, A.: A Chebyshev method for the solution of boundary value problems. J. Comput. Phys. **53**, 443–455 (1984)
65. Zebib, A.: Removal of spurious modes encountered in solving stability problems by spectral methods. J. Comput. Phys. **70**, 521–525 (1987)

Chapter 3
The Chebyshev Collocation Method

Abstract The chapter is devoted to the efficient implementation of Chebyshev collocation method. First, the performances of the method in solving fourth order GEPs are compared with those of ChT and ChG counterparts. Then, ChC method is used to solve some genuinely high order, i.e., larger than two, and/or singularly perturbed eigenvalue problems. Two of them, of sixth and eighth order represent linear hydrodynamic stability problems. Also some fourth order problems with variable coefficients (tensile instabilities of thin annular plates etc.) are successfully considered. In order to reduce the high order problems to systems of second order equations supplied with Dirichlet boundary conditions we introduce a so called "$D^{(2)}$" *strategy* or factorization. Using this strategy with $N = 2^{10}$ a conjecture with respect to the first eigenvalue of the Viola's problem is stated. This is a fourth order singularly perturbed eigenvalue problem. A special attention is paid to the well known Mathieu's system as a MEP. A lot of eigenmodes and eigenfrequencies corresponding to various geometries of the vibrating elliptic membrane problem, in which this system is originated, are displayed. In order to avoid spurious eigenvalues (at infinity) and to improve the computation of a specified region of the spectrum, mainly in case of large problems, some Jacobi Davidson type methods are used. Making use of the pseudospectrum of a singular GEP we comment on the backward stability and the order of convergence of *JD* and Arnoldi methods in computing the first two eigenvalues.

Keywords Chebyshev collocation · Jacobi Davidson method · Linear hydrodynamic stability · Mathieu's multiparameter problem · Phase space differentiation · Viola's eigenvalue problem

> *It all sounds too good to be true. Our theme is that actually, the collocation method is as good as advertised but only if one is a little bit careful. It is a truism in auto safety that most accidents occur within 10 miles of home. Its* arithmurgical *equivalent is that no calculation is too simple to screw up.* J. P. Boyd, [3]

C.-I. Gheorghiu, *Spectral Methods for Non-Standard Eigenvalue Problems*,
SpringerBriefs in Mathematics, DOI: 10.1007/978-3-319-06230-3_3,
© The Author(s) 2014

3.1 ChC Method Versus ChG and ChT Methods in Solving Fourth Order GEPs

The collocation method is more reliable than deriving a banded Galerkin matrix by means of recurrence relations; the pseudospectral code is simple to check, whereas it is easy to make an intestable mistake with the intricate algebra required for the Galerkin method. J. P. Boyd [3]

In order to compare the performances of spectral methods as well as the properties of their differentiation matrices in solving fourth order problems we will consider in turn some examples.

Buckling column with clamped boundary conditions-revisited The first one refers to Eq. (1.9) supplied with *clamped boundary conditions* (2.13). The problem (1.9)–(2.13) is again rewritten in $(-1, 1)$ and is first discretized by some ChG methods. The simplicity of boundary conditions and some suggestion provided by the weight factors (1.66) introduced in Sect. 1.2 enables the construction of various test and trial functions. Therefore, along with Shen's basis (see Lemma 2.2) and Heinrichs' basis (see (2.15) and Lemma 2.3) one can imagine bases of the form $(1 - x^2)(T_k(x) - T_{k-2}(x))$, for suitable k, etc.

However, the ChGS formulation implies the GEP

$$- E\mathbf{u} = \lambda D\mathbf{u}, \tag{3.1}$$

with the matrices E and D defined in Lemma 2.2 and ChGHS formulation implies the GEP

$$-GD^4\mathbf{u} = \lambda GD^2\mathbf{u}, \tag{3.2}$$

where the matrices GD^4 and GD^2 are defined in Lemma 2.3. In both problems $\mathbf{u} = (u_0, \ldots, u_N)^T$ stands for the eigenvector.

In ChC the boundary conditions at ends are introduced by a *removing technique of independent boundary conditions* introduced in [27] (see Sect. 4.4).

Eventually ChT method means to solve the algebraic system (2.11–2.14).

We mention that exact value of the first eigenvalue (the buckling load) is $\lambda_{ex} = (2\pi)^2$. The closed form of the first eigenvector (the shape of the buckling column) is $\phi_1(x) = 1 - \cos(2\pi x)$ (see [5]).

Inspecting Fig. 3.1 it is fairly clear that Galerkin type methods perform better than collocation method irrespective of cut off parameter N. Both methods surpass the tau method. However, at N around 2^6 the accuracy of all four of them is comparable. The price to be paid in using Galerkin type methods consists in the necessity to build up test and trial functions which implicitly satisfy all *essential* boundary conditions.

The performances of Galerkin type methods are partly justified by their pseudospectra and Henrici's number.

Comparing Figs. 3.2, 3.3 and 3.4 it is apparent that the differentiation matrices provided by Shen's method are the closest to symmetry. This conclusion is confirmed by the entries in the Table 3.1.

Fig. 3.1 $log_{10} |\lambda_1 - \lambda_{ex}|$ when the problem (1.9)–(2.13) is solved by ChC method (*starred line*), ChGS method (*circled line*), ChGHS method (*dotted line*) and crude ChT method (*crossed line*)

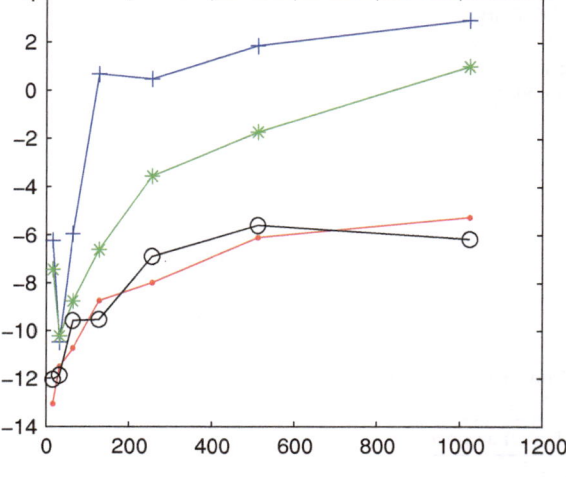

Fig. 3.2 Pseudospectrum of ChGS fourth order (*left*) and second order differentiation matrices (see Lemma 2.2)

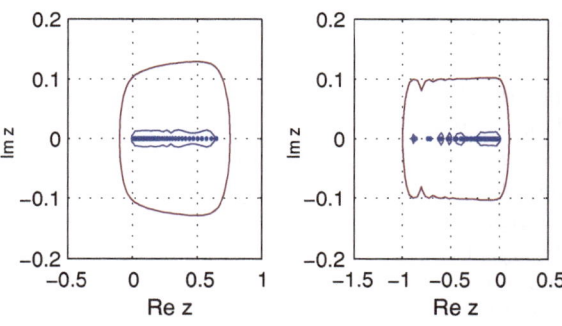

Fig. 3.3 Pseudospectrum of ChGHS fourth order (*left*) and second order differentiation matrices (see Lemma 2.3)

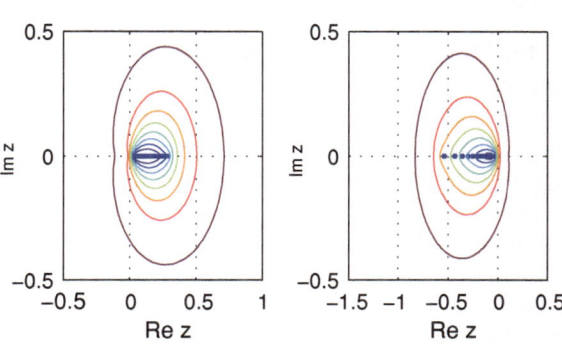

With a similar strategy can be solved the problem (1.10).

Remark 3.1 In the handbook chapter [1] the authors provide a lot of variational formulations for one and two dimensional eigenvalue problems. At the same time

Fig. 3.4 Pseudospectrum of
ChC fourth order (*left*) and
second order differentiation
matrices with clamped
boundary conditions

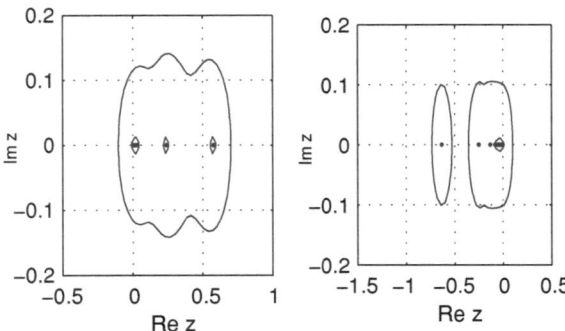

Table 3.1 Scalar measures for second and fourth order differentiation matrices based on Chebyshev polynomials

Henrici numbers	ChC	ChT
$D^{(2)}$	0.2027	0.9278
$D^{(4)}$	0.8864	1.0043

they expose a rigorous functional framework for the existence and multiplicity of their solutions. The numerical approximations of solutions have been obtained by *local methods* such as finite differences and various finite elements schemes. The authors observe that "because the rate of convergence for eigenvalues is twice that for eigenfunctions, we see that the eigenvalues are much cheaper to compute than the eigenfunctions". With respect to ChC method, Fig. 3.7 enables us to state that the rate of convergence for the first eigenvalue is at least twice that for the first eigenvector for $N = O(2^7)$. However, the additional price to be paid to compute the first eigenvector is insignificant.

Buckling column with a sort of mixed boundary conditions-revisited We consider now the problem of buckling load but supplied with mixed boundary conditions, i.e., (1.9)–(2.1). The pseudospectrum of the ChC matrix with enforced boundary condition is depicted in Fig. 3.5. It looks "smoother" than that of ChT corresponding matrix reported in Fig. 2.1 The Henrici numbers are reported in Table 3.2.

As the Galerkin methods are difficult to be applicable due the mixed boundary conditions ChC seems to be the better choice. We have obtained a fairly good approximation for the first eigenvalue. This approximatively equals 4.4934^2 (see for instance [5, p. 793]). The problem is easy solvable with ChT method as well. Thus, we have set up a MATLAB code using differentiation formulas (1.25) and their counterparts for boundary values (1.26). Actually we have solved the algebraic GEP (2.2)–(2.11). A fairly reasonable approximation for the first eigenvalue was obtained.

The first three eigenmodes obtained by ChC are depicted in Fig. 3.6. They exactly satisfy all boundary conditions. Moreover, the first computed eigenmode provides a correct shape of the buckling column for first load λ_1. Unfortunately, the first mode

Fig. 3.5 Pseudospectrum of ChC fourth order differentiation matrix with mixed boundary conditions

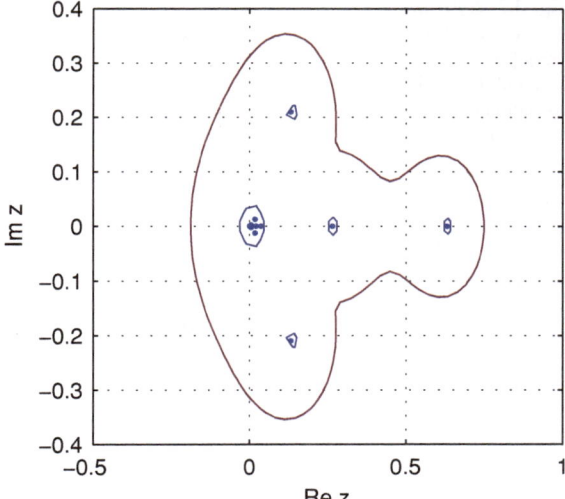

Table 3.2 Scalar measures for second and fourth order differentiation matrices corresponding to problem (1.9)–(2.1)

Henrici numbers	ChGS	ChGHS	ChC	ChT
$D^{(2)}$	0.0332	0.3131	0.2329	0.8344
$D^{(4)}$	0.1672	0.4018	0.5593	1.0018

Fig. 3.6 The first three eigenmodes for the problem (1.9)–(2.1). The *star line* stands for the first eigenmode in closed form

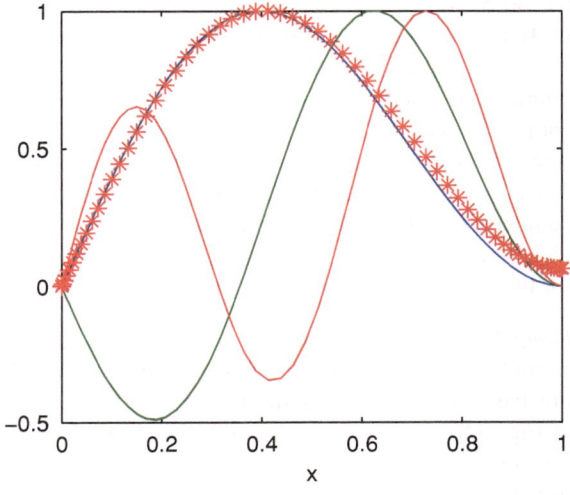

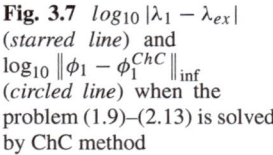

Fig. 3.7 $log_{10} |\lambda_1 - \lambda_{ex}|$ (*starred line*) and $log_{10} \|\phi_1 - \phi_1^{ChC}\|_{inf}$ (*circled line*) when the problem (1.9)–(2.13) is solved by ChC method

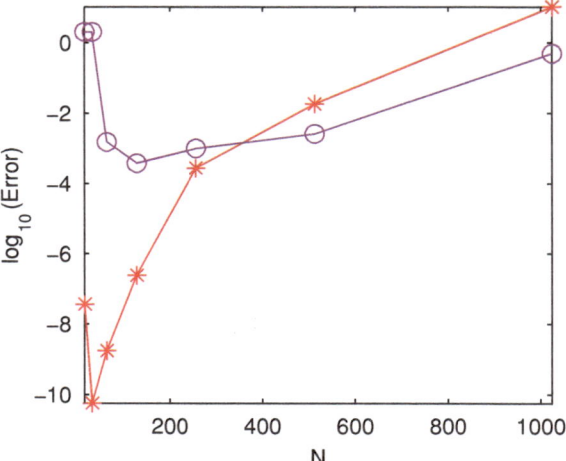

Table 3.3 Scalar measures for second and fourth order differentiation matrices corresponding to problem (1.9)–(2.10)

Henrici numbers	ChC	ChT
$D^{(2)}$	0.2273	0.9645
$D^{(4)}$	0.7215	1.0097

provided in the textbook above quoted, i.e., $\phi_1(x) = sin\sqrt{\lambda_1} - x\sqrt{\lambda_1}cos\sqrt{\lambda_1}$, satisfies only approximatively the boundary conditions in the right end.

Buckling column with hinged boundary conditions-revisited We consider now the problem of buckling load but supplied with hinged boundary conditions, i.e., (1.9)–(2.10). The Henrici numbers for thus situation are reported in Table 3.3.

As the Galerkin methods are again difficult to be applicable due the lacunar interpolation involved in the hinged boundary conditions ChC seems to be the better choice. We have obtained a fairly good approximation for the first eigenvalue $\lambda_1 = \pi^2$ (see for instance [5, p. 793]). The problem is easy solvable with ChT method as well. Thus, we have set up a MATLAB code using differentiation formulas (1.25) and their counterparts for boundary values (1.26). Actually we have solved the algebraic GEP (2.11)–(2.12). A fairly reasonable approximation, i.e., with first five decimal digits, for the first eigenvalue was obtained.

Remark 3.2 We have observed so far that the Heinrichs' basis (2.15) was useful in formulating ChT method and ChG method as well. It is very important to remark that this basis is even more useful in setting the ChC method. The situation is explained in Fig. 3.8. By far, the best conditioned are the collocation differentiation matrices corresponding to Heinrichs' basis when compared with usual ChC differentiation matrices or ChC differentiation matrices with homogeneous Dirichlet boundary

Fig. 3.8 Log of the condition number of the matrices: $D^{(2)}$ (*square line*), $\widetilde{D}^{(2)}$ (*circle line*) and $D^{(2)}$ Heinrichs' basis (*diamond line*) left part, respectively $D^{(4)}$ (*star line*) and $D^{(4)}$ Heinrichs' basis (*triangle line*) right part

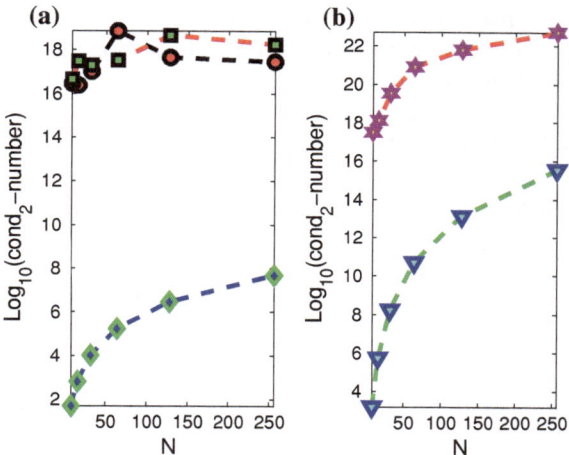

conditions enforced. The statement becomes even more clear when the conditioning in Figs. 3.13 and 3.27 are taken into account. In order to find the differentiation matrices corresponding to Heinrichs' basis we have used the code polydif from [47]. It computes the differentiation matrices corresponding to a general set of interpolating polynomials for arbitrary sets of nodes and weights. In spite of this better conditioning, our numerical experiments carried out with ChC method based on Heinrichs' basis do not show a perceptible improving of the result obtained by standard ChC.

Remark 3.3 The GEP (1.9)–(2.1) can be cast into the following *variational* and *minimization* formulations respectively

$$\textit{find } u \in K \textit{ and real } \lambda \textit{ such that } \int_0^1 u''v''dx = \lambda \int_0^1 u'v'dx, \ \forall v \in K, \quad (3.3)$$

and

$$\textit{find } \lambda = \min_{u \in K \backslash \{0\}} \frac{\int_0^1 \left(u''\right)^2 dx}{\int_0^1 (u')^2 dx}, \quad (3.4)$$

where K is the closed subspace of $H^4(0, 1)$ defined by

$$K = \left\{ v \in H^4(0, 1) \mid v(0) = v''(0) = v(1) = v'(1) = 0 \right\}. \quad (3.5)$$

The minimization formulation assures that the first eigenvalue is positive. We have to underline that the same equation supplied with the clamped boundary conditions (2.13) leads to formally identical weak and respectively minimization formulations. The unique difference consists in the definition of the set K. In case of clamped boundary conditions, K equals $H_0^2(0, 1)$.

Tensile instabilities of thin annular plates The tensile instabilities of thin annular plates are studied in [7] and consist in solving the following fourth order eigenvalue problem with *variable coefficients*

$$W'''' + \mathscr{A}(\rho, \Lambda)\, W''' + \mathscr{B}(\rho, \Lambda)\, W'' + \mathscr{C}(\rho, \Lambda)\, W' + \mathscr{D}(\rho, \Lambda)\, W = 0, \ \eta < \rho < 1, \tag{3.6}$$

where

$$\mathscr{A}(\rho, \Lambda) := \frac{2}{\rho}, \quad \mathscr{B}(\rho, \Lambda) := -\left[\frac{2n^2 + 1}{\rho^2} + \Lambda^2\left(-1 + \frac{1}{\rho^2}\right)\right], \tag{3.7}$$

$$\mathscr{C}(\rho, \Lambda) := \frac{1}{\rho}\left[\frac{2n^2 + 1}{\rho^2} + \Lambda^2\left(1 + \frac{1}{\rho^2}\right)\right],$$

$$\mathscr{D}(\rho, \Lambda) := \frac{n^2}{\rho^2}\left[\frac{n^2 - 4}{\rho^2} - \Lambda^2\left(1 + \frac{1}{\rho^2}\right)\right]. \tag{3.8}$$

In order to determine the eigenmode $W(\rho, n, \eta)$ and the eigenvalues $\Lambda(n, \eta)$ the following set of mixed boundary conditions is attached

$$W = 0, \rho = \eta,$$

$$W'' + \frac{\nu}{\rho}W' - \frac{\nu n^2}{\rho^2}W = 0, \ \rho = \eta, 1,$$

$$W''' + \frac{1}{\rho}W'' - \left[\frac{1 - (2 - \nu)n^2}{\rho^2}\right]W' + \left[\frac{(3 - \nu)n^2}{\rho^3}\right]W = 0, \rho = 1, \tag{3.9}$$

where ν is a material constant which may have any value between 0 and 0.5, but usually is about 0.3 and $\eta := a/b$ with a the inner radius and b the outer radius of a thin annular plate. The main aim is to discuss the dependence of Λ on a/b and n, which stands for the *mode number* in the solution of the bifurcation equation. In [7] the authors use the *compound matrix method* in order to solve the problem (3.6)–(3.9). They claim that it is a well tested technique which performs particularly well in computing the response curves for a range of problems in solid mechanics.

Due to the presence of the variable coefficients, and partly due to the high order of derivatives in boundary conditions, we consider that tau and Galerkin methods have slender chances to be used in order to solve the problem. Instead, ChC method successfully solved the (3.6)–(3.9) GEP. The results, depicted in Fig. 3.9, qualitatively agree with those reported in [7].

Another fourth order eigenvalue problem with variable coefficients In [6] the following fourth order problem supplied with hinged boundary conditions is solved by *extended sampling method*:

Fig. 3.9 The response curves $\Lambda = \Lambda(\eta, n)$ for buckling problem (3.6)–(3.9) corresponding to $n = 3, 8, 13$

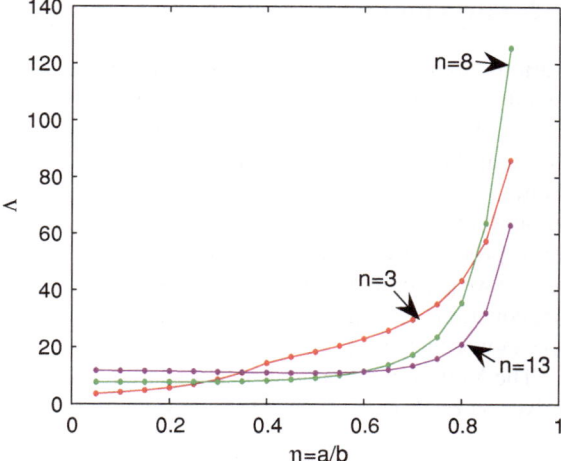

Fig. 3.10 The first four eigenvectors of problem (3.10)

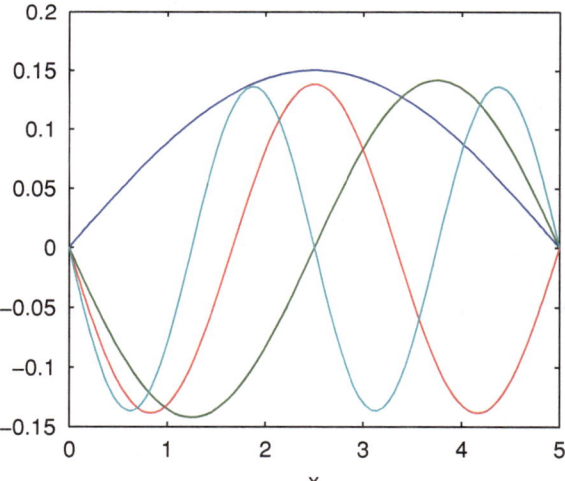

$$\begin{cases} u'''' - 0.02x^2u'' - 0.04xu' + \left(0.0001x^4 - 0.02\right)u = \lambda u, \ 0 < x < 5, \\ u(0) = u''(0) = u(5) = u''(5) = 0. \end{cases} \quad (3.10)$$

The first four eigenvalues computed by ChC have the following numerical values: $1.44914725e-001$, $2.49070294e+000$, $1.26345219e+001$ and $3.99276550e+001$. The corresponding eigenvectors are depicted in Fig. 3.10. They have unitary norm and can be distinguished because each eigenmode u_i, $i = 1, 2, 3, 4$ has $i - 1$ roots. Our numerical results agree to some extent with the results reported in [6]. The author of this paper claims that his results are verified with some other results obtained using the code SLEUTH (see [22]). As these latter results are not visible we can not comment more on their accuracy.

3.2 The Viola's Eigenvalue Problem

Various high order eigenvalue problems were solved by spectral methods using mainly tau and collocation variants. In this respect we mention the following papers [2, 4, 18, 24, 28, 39, 43] to quote but a few. For one-dimensional fourth order problems in [13] the authors introduce a factorization which shell prove fairly useful from now on. However, as we do not have theoretical estimations for the spectrum of our high order hydrodynamic stability problems we will use the following eigenvalue problem as a benchmark one. More than that, we try to make more explicit the reason we prefer to transform high order eigenvalue problems in systems of differential equations containing only second order derivatives, i.e., the so called the "$D^{(2)}$" *strategy* or factorization.

The Viola's eigenvalue problem is a fairly challenging one, still not completely solved. It reads (see [38, p. 218] and some works quoted there)

$$\begin{cases} D^2[(1-\theta x)^3 D^2 u] - \lambda (1-\theta x) u = 0, & x \in (0,1), \\ u = D^2 u = 0, & x = 0, 1, \end{cases} \tag{3.11}$$

where the parameter θ satisfies $0 \le \theta < 1$ and $D := \frac{d}{dx}$. The problem becomes singular as $\theta \to 1^-$. By straightforward variational arguments (see for instance [23, Chap. 10] for an easy introduction) it can be rewritten as a minimization one, namely

$$\lambda = \min_{u \in K \setminus \{0\}} \frac{\int_0^1 (1-\theta x)^3 \left(D^2 u\right)^2 dx}{\int_0^1 (1-\theta x) u^2 dx},$$

where K is the closed subspace of $H^4 (0,1)$ defined by

$$K := \left\{ v \in H^4 (0,1) \mid v(0) = D^2 v(0) = v(1) = D^2 v(1) = 0 \right\}.$$

$D^{(2)}$ **strategy** With a change of variable

$$v := (1-\theta x)^3 D^2 u,$$

the problem (3.11) reduces to an eigenvalue problem for the pair $(u \ v)^T$ supplied only with Dirichlet boundary conditions, namely

$$\begin{cases} (1-\theta x)^3 D^2 u - v = 0, & x \in (0,1), \\ D^2 v = \lambda (1-\theta x) u, \\ u = v = 0, & x = 0, 1. \end{cases}$$

As we make use of the ChC scheme, the linear transformation of independent variables $z := 2x - 1$ shifts our problem on the close interval $[-1, 1]$. The method casts the problem into the following singular algebraic GEP

$$Aw = \lambda Bw, \tag{3.12}$$

where the block matrices A and B are

$$A = \begin{pmatrix} (Dx)^3 \cdot \widetilde{D}^{(2)} & -I \\ Z & \widetilde{D}^{(2)} \end{pmatrix}, \quad B = \begin{pmatrix} Z & Z \\ Dx & Z \end{pmatrix}, \tag{3.13}$$

and the sub matrix Dx of order $N - 1$ is defined $Dx := \mathbf{diag}\,(1 - \theta\,(\mathbf{x}_{int} + 1)\,/2)$.

Recall that the matrix $\widetilde{D}^{(2)}$ has been obtained by enforcing the Dirichlet boundary conditions in the second order differentiation matrix computed on the whole set of $N + 1$ (CGaussL) nodes. Thus the rows and columns corresponding to the first node x_0 and the last node x_N are deleted in $D^{(2)}$. The vector $\mathbf{w}$ contains the values of the approximations of u and respectively v in the nodes of $\mathbf{x}_{int}$, i.e., $\mathbf{w} := (u_1, \ldots, u_{N-1}, v_1, \ldots, v_{N-1})^T$.

The problem (3.12) was solved using the MATLAB code `eig`. Our results confirm up to some extent those reported in [10, 38]. Moreover, we have pushed the computations to extreme values of parameter θ and order of ChC approximation. Thus, for $\theta = 1 - (1.0e - 6)$ and $N = 1{,}024$ we have obtained

$$\lambda_1 = 1.05351026442501, \quad \lambda_2 = 130.714326912961.$$

Consequently, we formulate the following

Conjecture 3.1 The first eigenvalue λ_1 of the problem (3.11) approaches 1 as $\theta \to 1^-$!

This statement is additionally supported by the numerical results reported in Fig. 3.11. They were obtained when $N = 512$ and θ in the range $[0.5, 1 - (1.0e - 06)]$. It is fairly clear that the first eigenvalue decreases very sharply as $\theta \to 1^-$.

An inconvenience of our factorization consists in the fact that corresponding to all zero rows in the matrix B (the rank of matrix B equals $N - 1$ and that of A in (3.13) is double) the QZ algorithm provides eigenvalues at infinity.

As a further check on accuracy, we have computed two types of sensitivities defined as

$$S^{(i)} := \frac{\|\mathbf{x}_i\|\,\|\mathbf{y}_i\|}{\mathbf{x}_i^H \mathbf{y}_i}, \quad S1^{(i)} := \frac{\|\mathbf{x}_i\|\,\|\mathbf{y}_i\|}{\sqrt{\left|\mathbf{y}_i^H A\mathbf{x}_i\right|^2 + \left|\mathbf{y}_i^H B\mathbf{x}_i\right|^2}}, \tag{3.14}$$

for the ith eigenvalue where $\mathbf{x}_i$ and $\mathbf{y}_i$ are the right and left eigenvectors of (3.12) respectively. They were introduced in [40] and used in an analogous problem in [8]. Corresponding to ChC method applied to problem (3.12) with $N = 1{,}024$ we summarize the situation in the Table 3.4.

It is obvious from this table that both measures of sensitivity drastically increase along with the parameter θ. In other words, the more singular is the differential problem, the more sensitive are the eigenvalues. Since the quantity $\log_{10} S^{(i)}$ is a measure

Fig. 3.11 The behavior
of the eigenvalue λ_1 as θ
approaches 1^-

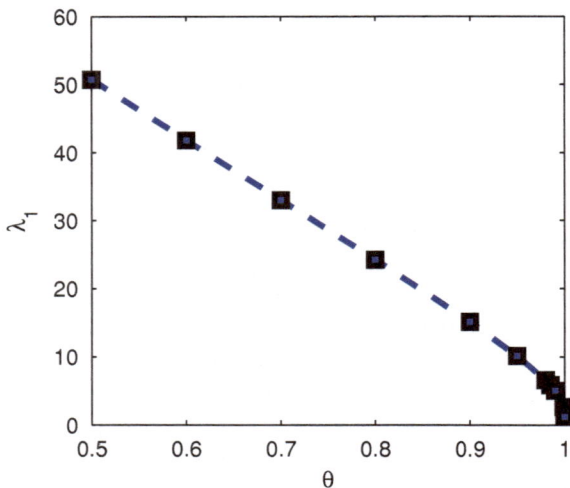

Table 3.4 The first two eigenvalues and their sensitivities

θ	λ_1	$S^{(1)}$	$S1^{(1)}$	λ_2	$S^{(2)}$	$S1^{(2)}$
0.5	50.7162236810491	1.88	1.44	838.20892567242	7.66	1.55
$1 - (1.0e - 6)$	1.05351026442501	56.42	$16.11e + 05$	130.714326912961	42.54	$16.11e + 05$

of the number of digits one lose in accuracy this table provides useful information
on the accuracy of the actual computations. A similar conclusion provides Fig. 3.12.
As the curves $\log_{10} cond(A)$ versus θ, for $N = 256$ and $N = 512$ respectively, are
quite parallel and horizontal for $0.6 \le \theta < 0.95$ they sharply increase for $\theta \to 1^-$.

However, the Henrici's number for matrix A equals 0.3573 and is totally independent on θ. This value is in accordance with the usual values for ChC method (see
Table 3.2).

Remark 3.4 An attempt was also made to solve the problem (3.11) as a fourth order
one. In order to fulfill the boundary conditions we have built this time the ChC method
on the more complicated weighted functions, namely

$$\left(1 - x^2\right)^3 T_i(x), \quad i = 0, 1, 2, \ldots, N. \tag{3.15}$$

To obtain the second and fourth order differentiation matrices we have appealed to the
code polydif from the same suite [47]. In this situation the right hand side matrix
of the generalized eigenvalue problem is a diagonal one and consequently it is easy
invertible. This way we are in the position to solve an usual eigenvalue problem.
For values of N around 128, the MATLAB code eigs as well as the MATLAB
code invshift (see [32, p. 175]), which evaluates the eigenvalue of minimum

Fig. 3.12 The conditioning of matrix A with respect to N and θ

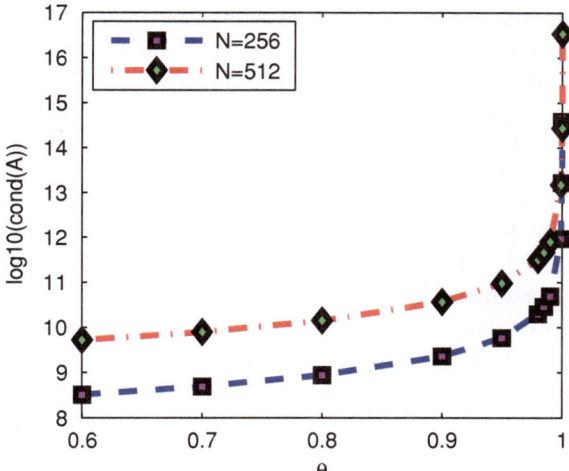

modulus of a matrix by the inverse power method, still provide positive values for λ_1. However, they are far larger than those obtained above. For N beyond 128 the situation is even worse. The eigenvalue of minimum modulus becomes negative and also complex eigenvalues appear in the superior part of the spectrum. This means *numerical instability*. In order to explain this phenomenon, we observe that the Henrici's number for fourth order differentiation matrix based on (3.15) equals 0.7135. This value is considerably larger than the usual values for ChC method (see Table 3.1). The conditioning of this matrix proved to be even much worse than that reported in Fig. 3.12. Thus, we assess that this lack of stability is a consequence of a cumulative effect of bad conditioning and non-normality of the fourth order derivative matrix corresponding to the above weighted functions.

Remark 3.5 In [10] the author uses the orthogonal invariants method in order to solve the problem. Our numerical results confirm the bounds provided in this paper for $\theta = 0.5$ and the results from [38]. In fact, for $\theta = 0$ the Viola's problem reduces to the problem of transverse vibrations of a uniform elastic bar supplied with hinged boundary conditions, i.e., (1.8)–(2.10). The exact eigenvalues are $\lambda_1 = \pi^4$ and $\lambda_2 = (2\pi)^4$. Our $D^{(2)}$ strategy confirms these values with spectral accuracy. It is important to observe that with the above computations of the firs two eigenvalues of Viola's problem we correct some typos from our paper [15].

3.3 Linear Hydrodynamic Stability of Thermal Convection with Variable Gravity Field

Variations in the gravity field occurring below, on and above the Earth's surface have been intensively studied. Usually, on a laboratory scale these variations are ignored and the gravity field can be considered as constant. However, their presence in crystal growth, in convection problems from porous media, and other convective motions from biomechanics, chemistry or geophysics makes them attractive from the applications point of view. Consequently, their effects on the stability bounds must be investigated.

In the present section we are concerned with a problem governing both the convection and conduction in a horizontal layer of a fluid heated from below and bounded by two impermeable walls. Across the layer the gravity field will vary according to some specific laws. The heat conducting viscous fluid is contained in a layer situated between the planes $z = 0$ and $z = h$. For $t > 0$ the conduction and convective motion is governed by the conservation equations of momentum, mass and internal energy ([38, p. 93])

$$
\begin{cases}
\dfrac{\partial \mathbf{v}}{\partial t} + (\mathbf{v} \cdot \mathbf{grad})\mathbf{v} = -\dfrac{1}{\rho}\mathbf{grad}\,p + \nu\Delta\mathbf{v} + \mathbf{g}(z)\alpha T, \\
\mathbf{div}\,\mathbf{v} = 0, \\
\dfrac{\partial T}{\partial t} + (\mathbf{v} \cdot \mathbf{grad})T = k\Delta T,
\end{cases}
\tag{3.16}
$$

where, as usual, ν is the coefficient of kinematic viscosity, ρ is the density, α the thermal expansion coefficient, k the thermal diffusivity, p the pressure, T the temperature, $\mathbf{v}$ the velocity field and $\mathbf{g}(z) := g(1 + \varepsilon h(z))\mathbf{k}$ is the variable gravity field, with the gravity g constant and $\mathbf{k}$ the unit vector in the z-direction.

The linear stability of the conduction stationary solution of equations (3.16), against normal mode perturbations, written in the non-dimensional form, and corresponding to *free boundaries* is governed by the following two-point boundary value problem

$$
\begin{cases}
(D^2 - a^2)^2 W = R(1 + \varepsilon h(z))a^2\Theta, \\
(D^2 - a^2)\Theta = -RW,
\end{cases}
\tag{3.17}
$$

$$
W = D^2 W = \Theta = 0, \quad z = 0, 1.
\tag{3.18}
$$

The usual boundary conditions, i.e., for fixed rigid walls, read

$$
W = DW = \Theta = 0, \quad z = 0, 1.
\tag{3.19}
$$

This case is considered in detail in the monograph [38]. Up to our knowledge the case of free boundaries was not considered numerically.

Here $D := \frac{d}{dz}$, R^2 stands the Rayleigh number Ra and it represents the eigenparameter of the problem (3.17, 3.18). The parameter a is the wave number, ε is a "small" scale parameter and W and Θ are the amplitudes of the vertical velocity and respectively temperature perturbation. The pair $(W\ \Theta)^T$ form the eigenfunction of the eigenvalue problem.

We will consider only the free boundaries case. In [38] a numerical evaluation of the Rayleigh number in the case of rigid boundaries, by using the energy method, was also provided.

Taking into account the order of differentiation in (3.17, 3.18) we introduce the new variable

$$\Psi := (D^2 - a^2)W. \tag{3.20}$$

Thus, the two-point boundary value problem (3.17, 3.18) can be factorized as the second order system

$$\begin{cases} (D^2 - a^2)W - \Psi = 0, \\ (D^2 - a^2)\Psi - R(1 + \varepsilon h(z))a^2\Theta = 0, \\ (D^2 - a^2)\Theta + RW = 0, \end{cases} \tag{3.21}$$

supplied with the Dirichlet boundary conditions

$$W = \Psi = \Theta = 0, \ z = 0, 1. \tag{3.22}$$

Actually we have got a GEP.

It is worth noting at this moment that the eigenvalue problem (3.17, 3.18) can be rewritten more compactly in the following form. A sixth order differential equation

$$(D^2 - a^2)^3 W = -R^2(1 + \varepsilon h(z))a^2 W, \tag{3.23}$$

supplied with Dirichlet and *hinged* boundary conditions, namely

$$W = D^2 W = D^4 W = 0, \ z = 0, 1. \tag{3.24}$$

A *weak formulation* for the new problem (3.23, 3.24) reads
find $w \in K \setminus \{0\}$ and real Ra such that

$$\int_0^1 D^3 w D^3 v dz + 3a^2 \int_0^1 D^2 w D^2 v dz + 3a^4 \int_0^1 Dw Dv dz + a^6 \int_0^1 wv dz \\ = a^2 Ra \int_0^1 (1 + \varepsilon h(z))wv dz, \ \forall v \in K, \tag{3.25}$$

where the set K is a closed subspace of $H^6 (0, 1)$ [see the order of differentiation in (3.23)] defined by

$$K := \{v \in H^6 (0, 1) \,|\, v(0) = v(1) = D^2 v(0) = D^2 v(1) = D^4 v(0) = D^4 v(1) = 0\}.$$

More than that, the critical Rayleigh number can be characterized by the following *minimization problem*

Fig. 3.13 Log of the
condition number of the
Chebyshev differentiation
matrices $\widetilde{D}^{(2)}$ (*circle
line*), $D^{(2)}$ (*square line*),
$D^{(4)}$ (*diamond line*),
$D^{(6)}$ (*hexagram line*) and
$D^{(8)}$, (*upward-pointing
triangle*) versus N

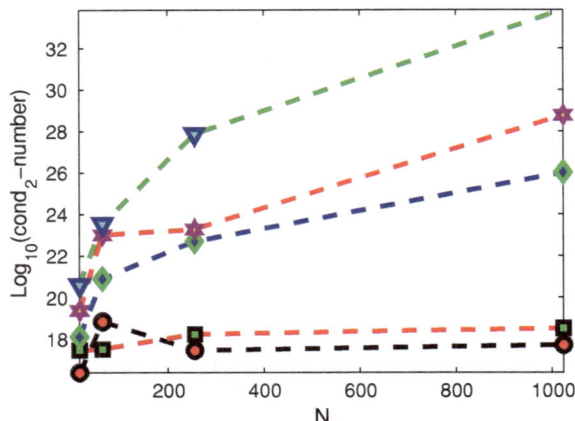

$$Ra = \min_{v \in K \setminus \{0\}} \frac{\int_0^1 (D^3 v)^2 dz + 3a^2 \int_0^1 (D^2 v)^2 dz + 3a^4 \int_0^1 (Dv)^2 dz + a^6 \int_0^1 v^2 dz}{a^2 \int_0^1 (1 + \varepsilon h(z)) v^2 dz},$$

(3.26)

whenever the quantity $1 + \varepsilon h(z)$ remains positive. In fact, in all our experiments this
condition is fulfilled.

It is fairly important to underline why we integrate numerically the problem (3.21,
3.22) instead of the problem (3.23, 3.24), its weak or minimization forms.

The main reason is the rapid increasing (deterioration) of the condition number
of the differentiation matrices as both the order of differentiation and the order of
approximation N get larger. The fact is illustrated in the Fig. 3.13. The second order
differentiation matrices $D^{(2)}$ and their counterparts $\widetilde{D}^{(2)}$, with the Dirichlet boundary
conditions enforced, have both a conditioning of order $O(N^6)$ for large N but the
latter behaves better.

Actually, we look for the smallest eigenvalue $Ra = R^2$ in the problem (3.21,
3.22) defining the neutral manifold. In our previous paper [15] we have solved this
problem by three different methods, a Galerkin one, a Petrov Galerkin one, both
based on Chebyshev polynomials and trigonometric functions, and by ChC using
"$D^{(2)}$" strategy.

The ChC method reduces the eigenvalue problem (3.21, 3.22) to a singular algebraic GEP of the form (3.12) where the matrices A and B are now

$$A = \begin{pmatrix} 4\widetilde{D}^{(2)} - a^2 I & -I & Z \\ Z & 4\widetilde{D}^{(2)} - a^2 I & Z \\ Z & Z & 4\widetilde{D}^{(2)} - a^2 I \end{pmatrix}, \quad B = \begin{pmatrix} Z & Z & Z \\ Z & Z & I - \varepsilon Dx \\ -I & Z & Z \end{pmatrix},$$

(3.27)

and Dx is the diagonal matrix $Dx := \mathbf{diag}\left((\mathbf{x}_{int} + 1)/2\right).$

Fig. 3.14 Neutral curves for various values of the physical parameters

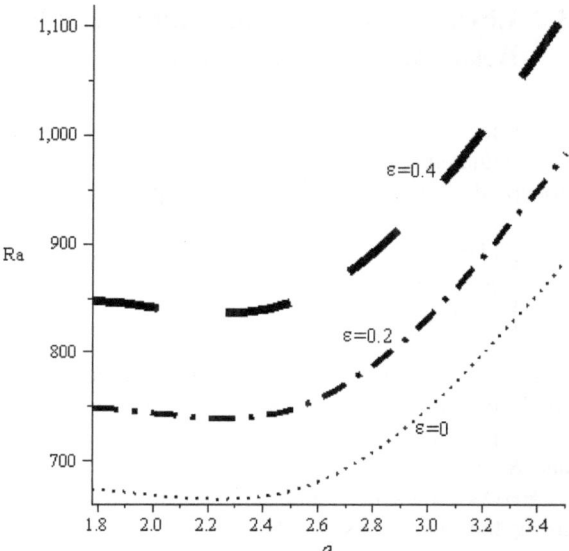

In order to find the entries in the matrices of algebraic eigenvalue problems corresponding to the first two methods we have used simple but tedious symbolic computations. These computations dominate all other costs of the methods and have been performed for N around 10. The numerical scheme for the generation of the spectrum in the case of Galerkin type methods based on trigonometric and polynomial expansions is a Maple code which implements the LU decomposition.

A better situation is observed with respect to the ChC method. The spectral order N used is much larger than that used in the weighted residual methods, i.e., N is of order 2^m with $5 \leq m \leq 10$. In spite of this, our numerical experiments showed that the collocation methods are much more rapid (the CPU time increases moderately with N) and more flexible with respect to the implementation process. However, both classes of methods provide reliable approximations for the lower part of the spectra (see Tables 1, 2 in our paper [15]).

More than that, they do not require assumptions on the *scale resolution*, i.e., the ratio Ra/N^2 is "small", as it is the case with the methods considered in [30]. It signifies a sort of reliability. With respect to the sensitivity defined by (3.14) we remark that this problem is less sensitive than the singularly perturbed Viola's problem. For the first indicator we obtained $S^{(1)} = 5.789210$ which increases to $S^{(2)} = 15.846905$ and $S1^{(i)}$, $i = 1, 2$ slightly oscillates around 0.59. It is important that they remain practically independent of N.

In Fig. 3.14 graphical representations of the neutral curves for different values of the scale parameter are given, pointing out its influence on the stability domain.

3.4 Linear Hydrodynamic Stability of EHD Convection Between Two Parallel Walls

The linear stability of the stationary solution in an EHD convection model, in a layer situated between the walls $z = \pm 0.5$, against normal mode perturbations, is governed by the following eigenvalue problem (see [14])

$$
\begin{cases}
(D^2 - a^2)^4 F - a^4 L F + a^2 R (D^2 - a^2) F = 0, & z \in (-0.5, 0.5), \\
F = D^2 F = D^4 F = D^6 F = 0, \\
z = \pm 0.5.
\end{cases}
\tag{3.28}
$$

Here F, which represents the amplitude of the temperature field perturbation, stands for the eigenfunction in (3.28). The physical parameter a represents the wave number, L is a parameter effectively measuring the potential difference between the planes and R stands for the Rayleigh number.

EHD systems have important industrial application in the construction of devices using the electro viscous effect or charge entrainment, for instance EHD clutch development and EHD high voltage generators.

The linear hydrodynamic stability of their steady states typically leads to high order differential eigenvalue problems.

A first important remark is in order at this moment. By standard variational arguments (integration by parts and imposing the boundary conditions) a *weak formulation* of the problem (3.28) can be obtained. It reads

find $F \in K$ and real Ra such that

$$
\int_{-\frac{1}{2}}^{\frac{1}{2}} \left(D^4 F \right) \left(D^4 V \right) dz + 2a^2 \left(2 + 3a^2 \right) \int_{-\frac{1}{2}}^{\frac{1}{2}} \left(D^3 F \right) \left(D^3 V \right) dz
$$

$$
+ 4a^6 \int_{-\frac{1}{2}}^{\frac{1}{2}} (DF)(DV) dz + a^4 \left(a^4 - L \right) \int_{-\frac{1}{2}}^{\frac{1}{2}} F V dz
$$

$$
= Ra^2 \left\{ \int_{-\frac{1}{2}}^{\frac{1}{2}} (DF)(DV) dz + a^2 \int_{-\frac{1}{2}}^{\frac{1}{2}} F V dz \right\}, \ \forall V \in K, \ I \equiv [-0.5, 0.5],
$$

$$
\tag{3.29}
$$

where K is the closed subspace of $H^8(I)$ defined by

$$
K := \left\{ v \in H^8(I) \mid v(\pm 0.5) = D^2 v(\pm 0.5) = D^4 v(\pm 0.5) = 0 \right\}.
$$

Moreover, whenever the parameters a and L satisfy the "ellipticity" condition, i.e.,

$$
a^4 - L \geq 0,
\tag{3.30}
$$

the problem is reduced to a *minimization* one, i.e.,

$$R = \inf_{u \in K \setminus \{0\}} \frac{N(u)}{DN(u)}, \qquad (3.31)$$

where

$$N(u) := \int_{-\frac{1}{2}}^{\frac{1}{2}} \left(D^4 u\right)^2 dz + 2a^2 \left(2 + 3a^2\right) \int_{-\frac{1}{2}}^{\frac{1}{2}} \left(D^3 u\right)^2 dz \qquad (3.32)$$

$$+ 4a^6 \int_{-\frac{1}{2}}^{\frac{1}{2}} (Du)^2 dz + a^4 \left(a^4 - L\right) \int_{-\frac{1}{2}}^{\frac{1}{2}} u^2 dz, \qquad (3.33)$$

and

$$DN(u) := a^2 \left\{ \int_{-\frac{1}{2}}^{\frac{1}{2}} (Du)^2 dz + a^2 \int_{-\frac{1}{2}}^{\frac{1}{2}} u^2 dz \right\}. \qquad (3.34)$$

It is clear that the smallest (algebraic) eigenvalue is real and positive, i.e., there exists a first $R > 0$ which satisfies (3.28).

In spite of the fact that in both formulations (3.29) and (3.31) the order of differentiation is halved when compared with that in (3.28), it remains too high in the perspective of numerical computations.

In order to solve numerically the problem (3.28) we introduce the new vector function

$$\mathbf{U} := (U_1, U_2, U_3, U_4)^T, \qquad (3.35)$$

with

$$U_1 := (D^2 - a^2)U_4, \ U_2 := (D^2 - a^2)U_1, \ U_3 := (D^2 - a^2)U_2, \ U_4 := F. \quad (3.36)$$

This substitution turns the eighth order equation from (3.28) into the following system of second order ordinary differential equations

$$\begin{cases} U_1 - (D^2 - a^2)U_4 = 0, \\ U_2 - (D^2 - a^2)U_1 = 0, \\ U_3 - (D^2 - a^2)U_2 = 0, \\ (D^2 - a^2)U_3 - a^4 L U_4 + a^2 R U_1 = 0. \end{cases} \qquad (3.37)$$

Consequently, the boundary conditions simplify considerably and become

$$U_1 = U_2 = U_3 = U_4 = 0, z = \pm 0.5. \qquad (3.38)$$

It is important to underline that, in this way, the high order (hinged) boundary conditions from the problem (3.28) are transformed into simple Dirichlet boundary conditions. This was also the successful strategy in our previous work [15].

The problem of linear stability in the EHD convection is an eighth order eigenvalue one with constant coefficients depending on two parameters, namely a and L. Consequently, the discussion of multiplicity of the roots of the characteristic equation when all the parameters are different from zero, becomes fairly laborious. In this context, the results obtained by direct analytical methods must be checked by alternative methods.

In this context, in our previous paper [9] we solved the problem (3.28) by direct (analytical) methods as well as by Galerkin methods based on Chebyshev and Legendre polynomials. A particular attention was given to ChC coupled with "$D^{(2)}$" strategy. With this latter method we cast the problem into a singular GEP, namely

$$A_1 \mathbf{X} = Ra^2 \, B_1 \mathbf{X}, \tag{3.39}$$

where the block matrices A_1 and B_1 are

$$A_1 := \begin{pmatrix} I & O & O & -4\widetilde{D}^{(2)} + a^2 I \\ -4\widetilde{D}^{(2)} + a^2 I & I & O & O \\ O & -4\widetilde{D}^{(2)} + a^2 I & I & O \\ O & O & 4\widetilde{D}^{(2)} - a^2 & -La^4 I \end{pmatrix}, \tag{3.40}$$

$$B_1 := \begin{pmatrix} O & O & O & O \\ O & O & O & O \\ O & O & O & O \\ -a^2 I & O & O & O \end{pmatrix}, \tag{3.41}$$

and the unknown vector is defined as

$$\mathbf{X} := (U_{1,1}, U_{1,2}, \dots U_{1,N-1}, U_{2,1}, \dots, U_{2,N-1}, U_{3,1}, \dots, U_{3,N-1}, U_{4,1}, \dots, U_{4,N-1})^T. \tag{3.42}$$

This way, the Dirichlet boundary conditions (3.38) are imposed, as usual, by deleting the first and the last row and column in the differentiation matrices. The MATLAB code `eig` was first used in order to solve the algebraic GEP (3.39). The value of the cut-off parameter $N = 16$ was large enough in order to assure the accuracy of the first eigenvalue. Numerical experiments with N between 2^6 and 2^8 do not improve the first eigenvalue. Numerical results obtained using the above factorization, along with analytical ones, and those obtained using Galerkin method are reported in Table 3.5. The numerical value $R_{c,ChC} = 657.1892$ corresponding to $a = \sqrt{4.92}$ and $L = 1$ has been validated in [41]. The author has used an operatorial ChT method.

Table 3.5 Numerical estimates of the *critical Rayleigh number* for various values of the parameters a, L obtained by LG, ChG and ChC in comparison with the analytical ones

a	L	$R_{c,analytical}$	$R_{c,LG}(N = 6)$	$R_{c,ChG}(N = 6)$	$R_{c,ChC}(2^4 \le N \le 2^8)$
2	0	667.0098	667.030	667.013	667.0092
$\sqrt{4.92}$	0	657.5133	657.543	657.527	657.5133
3	0	746.5276	746.54	746.530	746.5276
2	0	666.7421	666.742	666.725	666.7214
$\sqrt{4.92}$	0	657.1806	657.208	657.192	657.1892
3	0	746.0506	746.067	746.053	746.0506
2	10	664.1258	664.146	664.129	664.1258
$\sqrt{4.92}$	10	654.1866	654.193	654.177	654.1866
2	16	662.3954	662.399	662.416	35.8873

Eventually, we have to notice two important aspects with respect to our "$D^{(2)}$" strategy based on ChC method.

First, with accurately computed differentiation matrices, the method becomes extremely simply implementable in spite of the high order of differentiation. MATLAB concatenation is the recommended way. One of its drawback consists in the fact that when QZ algorithm is used it furnishes $3(N-1)$ spurious (at infinity) eigenvalues. They correspond to the $3(N-1)$ zero rows in the matrix B_1 [the rank of matrix B_1 equals $(N-1)$]. However, this fact does not affect the low part of the spectrum which is computed fairly accurate.

Second, we need to emphasize again that the applicability of Galerkin type methods heavily depends on the possibility to construct trial and test functions that satisfy all the *essential* boundary conditions of a given problem. They solve accurately high order non-standard eigenvalue problems or singularly perturbed ones but which are supplied with fairly simple boundary conditions. This is in fact the main disadvantage of these methods. Tau or collocation methods for problems involving complicated boundary conditions usually require less computational effort in order to enforce them.

3.5 Multiparameter Mathieu's Problem

Many of special functions have been vigorously studied over the past century. In contrast, the general case of multiparameter spectral theory has been rather neglected over the ears despite the fact that it arose almost as long ago as the classic work of Sturm and Liouville on one parameter eigenvalue problems. B. D. Sleeman [35]

Multiparameter spectral theory has its roots in solving boundary value problems for partial differential equations by classical method of separation of variables. In the standard case, the separation technique leads to the study of systems of ordinary differential equations linked by spectral parameters, i.e., separation constants, in an elementary way. Mathieu's system is often used in the literature as a motivating example for the introduction of MEPs, see, for example, the monograph [46]. This MEP approach of this well known system is the most natural one. The system is obtained if separation of variables is applied to solve the vibration of a fixed elliptic membrane. However, to the best of our knowledge, the accurate numerical solution of Mathieu's system as a two-parameter eigenvalue problem was never studied in detail. The two-parameter dependence makes computing Mathieu's functions more involved than, for example, Bessel's functions. In [34] the author provides some examples from science and technology arguing that they deserve accurate solutions of Mathieu's system.

MEPs of Mathieu's system The coupled system of Mathieu's angular and radial equations in which a and q are independent parameters will be thought of now as a multiparameter (two parameter) eigenvalue problem. The following four MEPs can be formulated with respect to Mathieu's system

- a π-even problem

$$
\begin{cases}
G''(\eta) + (a - 2q\cos 2\eta)\, G(\eta) = 0, & 0 < \eta < \pi/2, \\
\quad G'(0) = G'(\pi/2) = 0, \\
F''(\xi) - (a - 2q\cosh 2\xi)\, F(\xi) = 0, & 0 < \xi < \xi_0, \\
\quad F'(0) = F(\xi_0) = 0,
\end{cases}
\tag{3.43}
$$

- a 2π-even problem

$$
\begin{cases}
G''(\eta) + (a - 2q\cos 2\eta)\, G(\eta) = 0, & 0 < \eta < \pi/2, \\
\quad G'(0) = G(\pi/2) = 0, \\
F''(\xi) - (a - 2q\cosh 2\xi)\, F(\xi) = 0, & 0 < \xi < \xi_0, \\
\quad F'(0) = F(\xi_0) = 0,
\end{cases}
\tag{3.44}
$$

- a π-odd problem

$$
\begin{cases}
G''(\eta) + (a - 2q\cos 2\eta)\, G(\eta) = 0, & 0 < \eta < \pi/2, \\
\quad G(0) = G(\pi/2) = 0, \\
F''(\xi) - (a - 2q\cosh 2\xi)\, F(\xi) = 0, & 0 < \xi < \xi_0, \\
\quad F(0) = F(\xi_0) = 0,
\end{cases}
\tag{3.45}
$$

- a 2π-odd problem

$$
\begin{cases}
G''(\eta) + (a - 2q\cos 2\eta)\, G(\eta) = 0, & 0 < \eta < \pi/2, \\
\quad G(0) = G'(\pi/2) = 0, \\
F''(\xi) - (a - 2q\cosh 2\xi)\, F(\xi) = 0, & 0 < \xi < \xi_0, \\
\quad F(0) = F(\xi_0) = 0.
\end{cases}
\tag{3.46}
$$

These coupled systems of two-point boundary value problems come from the problem of a vibrating elliptic membrane Ω with fixed boundaries $\partial\Omega$,

$$\left(\Delta + \omega^2\right)\psi(x, y) = 0, \quad (x, y) \in \Omega, \quad \psi(x, y) = 0, \quad (x, y) \in \partial\Omega, \qquad (3.47)$$

when the separation of the variables, i.e., $\psi(x, y) := F(\xi)\,G(\eta)$ is used in the elliptical coordinates ξ and η,

$$\begin{aligned} x &:= h\cosh\xi\cos\eta, \\ y &:= h\sinh\xi\sin\eta, \quad 0 \leq \xi < +\infty, \quad 0 \leq \eta < 2\pi. \end{aligned} \qquad (3.48)$$

Thus

$$\xi_0 := \arccos\frac{\alpha}{h}, \qquad (3.49)$$

where $h := \sqrt{\alpha^2 - \beta^2}$ is half the distance between the foci of the membrane. The parameter q is related to the eigenfrequency ω by

$$q := \frac{h^2\omega^2}{4}, \qquad (3.50)$$

and a is the parameter arising in the separation of variables.

The above four systems are analyzed in details in the monograph [46]. For these *right definite* MEPs, the author provides results concerning the existence and countability of eigenvalues, numbers of zeros of eigenfunctions and the completeness of even and odd sets of eigenmodes. His analytical results are exhaustive. A Klein oscillation theorem is also available in [31] and some other useful comments on the formulations above can be found in [11]. The Mathieu's system can also be "embedded" in the most general setting of the MEP for ordinary differential equations formulated in [35].

ChC discretization Our previous numerical experiments concerning non-standard, high order and singularly perturbed eigenvalue problems proved that the ChC method is fairly accurate, reliable and implementable. It turned out to be sometimes superior to the spectral Galerkin or tau method, also based on the Chebyshev polynomials, for such problems. The well known monograph [12] provides even more hints with *how*, *when* and *why* this pseudospectral approach works.

Thus, the ChC discretization of MEP (3.43) reads

$$\begin{cases} \left(\left(\tfrac{4}{\pi}\right)^2 \cdot {}^{e,\pi}D_n^2 + (a - 2q \cdot \operatorname{diag}\left(\cos\pi\left(\mathbf{x}_{int} + 1\right)/2\right))\right)\mathbf{U} = 0, \\ \left(\left(\tfrac{2}{\xi_0}\right)^2 \cdot {}^{e,\pi}D_{nd}^2 - (a - 2q \cdot \operatorname{diag}\left(\cosh\xi_0\left(\mathbf{x}_{int} + 1\right)\right))\right)\mathbf{V} = 0, \end{cases} \qquad (3.51)$$

Fig. 3.15 The
overlapped pseudospectra of
problems (3.43) and (3.44),
$N = 24$, $\alpha = \cosh(2)$, $\beta =
\sinh(2)$

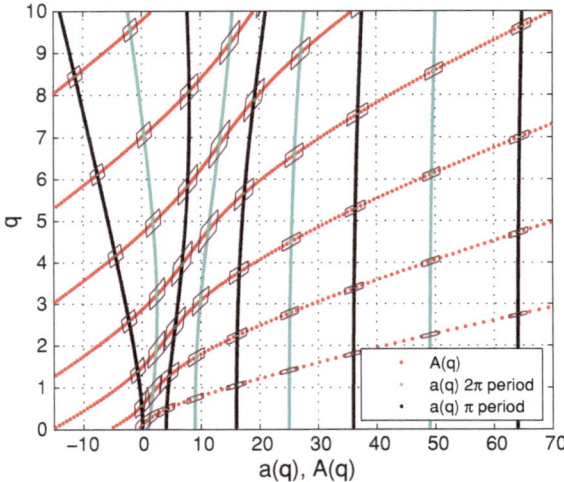

where $^{e,\pi} D_n^2$ and $^{e,\pi} D_{nd}^2$ are second order differentiation matrices in the (CGaussL) nodes. In the symbol $^{e,\pi} D_n^2$ the upper indices e and π stand for even and π period, and the lower index n for the Neumann boundary conditions

$$G'(0) = G'(\pi/2) = 0, \tag{3.52}$$

which are enforced. Similarly, in $^{e,\pi} D_{nd}^2$ the mixed boundary conditions

$$F'(0) = F(\xi_0) = 0, \tag{3.53}$$

are introduced, so n comes from the first and d from the second boundary condition respectively. We used the seminal paper [47] to obtain the entries of these two matrices and the general strategy from [27] to impose all boundary conditions. The vectors **U** and **V** contain the unknown values of G and F in the nodes (CGaussL).

Thus, the problem (3.51) is an algebraic MEP of type (A.1) with (a, q) standing for (λ, μ). The discretizations for the last three problems (3.44, 3.45) and (3.46) are analogous.

Unfortunately, the matrices $^{e,\pi} D_n^2$ and $^{e,\pi} D_{nd}^2$ are dense, non-symmetric and have high condition numbers (see for instance [15]). The pseudospectra (see [26] for numerical code) of even problems (3.43) and (3.44) are depicted in Fig. 3.15. This picture shows two mildly non-normal MEP with decreasing non-normality for large $a(q)$.

It is worth noting at this moment that the curves $a(q)$ represent the solutions of the first S-L problems in (3.43) *and* (3.44) for $q \in [0, 10]$. They are the interlaced quasi "vertical" curves in Fig. 3.15. The family of curves $A(q)$ depicts the solutions of the second S-L problem in (3.43) *or* (3.44) for the same range of q. They are

Fig. 3.16 The over-lapped pseudospectra of problems (3.45) and (3.46), $N = 24$, $\alpha = \cosh(2)$, $\beta = \sinh(2)$

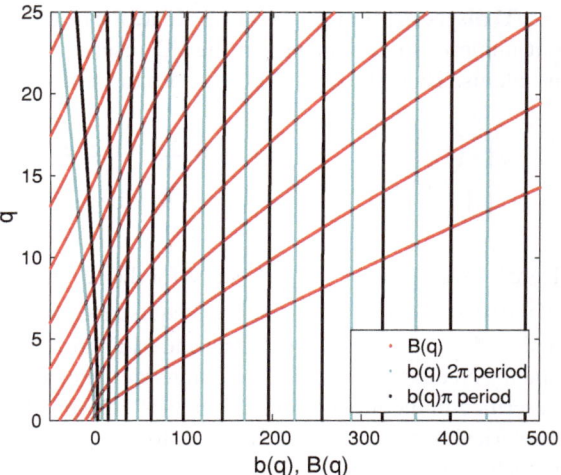

represented by the quasi "oblique" curves. Their intersections localize the eigenpairs $(a,\ q)$ of MEP (3.43). Our Fig. 3.15 refines Fig. 1 from the paper [31].

The ChC discretization of MEP (3.46) reads

$$
\begin{cases}
\left(\left(\frac{4}{\pi} \right)^2 \cdot {}^{o,2\pi} D_{dn}^2 + (a - 2q \cdot \mathrm{diag}\left(\cos \pi\ (\mathbf{x}_{int} + 1)\,/2 \right)) \right) U = 0, \\
\left(\left(\frac{2}{\xi_0} \right)^2 \cdot {}^{o,2\pi} D_d^2 - (a - 2q \cdot \mathrm{diag}\left(\cosh \xi_0\ (\mathbf{x}_{int} + 1) \right)) \right) V = 0.
\end{cases}
\tag{3.54}
$$

In the symbol ${}^{o,2\pi} D_{dn}^2$ the upper index o stands for odd property, the index 2π for period and dn for the mixed boundary conditions

$$
G\ (0) = G'\ (\pi/2) = 0.
\tag{3.55}
$$

The matrix ${}^{o,2\pi} D_d^2$ with the lower index d involves the symmetric Dirichlet boundary conditions

$$
F\ (0) = F\ (\xi_0) = 0.
\tag{3.56}
$$

The pseudospectra of odd problems (3.45) and (3.46) are depicted in Fig. 3.16. It shows, two even more normal problem than (3.43) and (3.44). The explanation consists in the fact that the Dirichlet boundary conditions

$$
F\ (0) = F\ (\xi_0) = 0
\tag{3.57}
$$

in (3.45) and (3.46) induce a sort of symmetry in the differentiation matrices, i.e., the matrix ${}^{o,2\pi} D_d^2$ is more normal than ${}^{e,\pi} D_{nd}^2 = {}^{e,2\pi} D_{nd}^2$.

FD discretization In order to evaluate the performances of our strategy we carried out numerical experiments on the FD discretization of our differential eigenvalue problems. Thus, the usual FD counterpart of (3.43) reads

$$
\begin{cases}
\left(\left(\frac{2}{\pi} \right)^2 \cdot {}^{e,\pi} D_n^{2,FD} + (a - 2q \cdot \mathrm{diag} \left(\cos \pi \left(X_{N+1} \right) \right)) \right) \mathbf{U} = 0, \\
\left(\left(\frac{1}{\xi_0} \right)^2 \cdot {}^{e,\pi} D_{nd}^{2,FD} - (a - 2q \cdot \mathrm{diag} \left(\cosh 2\xi_0 \left(X_N \right) \right)) \right) \mathbf{V} = 0.
\end{cases}
\tag{3.58}
$$

Here ${}^{e,\pi} D_n^{2,FD}$ and ${}^{e,\pi} D_{nd}^{2,FD}$ stand for the second order centered FD approximation of the second derivative in the $N+1$ equispaced nodes X_{N+1}. The matrices ${}^{e,\pi} D_n^{2,FD}$ and ${}^{e,\pi} D_{nd}^{2,FD}$ are now symmetric and tridiagonal of order $N+1$ and N respectively and the Neumann boundary conditions were introduced by *mirror imaging technique* described in the monograph [33, p. 549]. Despite these simplifications in (3.58) the numerical results provided further below are obviously inferior to those obtained by the ChC (see Tables 3.8, 3.9).

Numerical results It is important to point out at this moment that in their paper [50] the authors use a Fourier collocation method (and not Galerkin!) in order to discretize Mathieu's system. Their shape (trial) functions are some trigonometric functions which implicitly satisfy the boundary conditions but it is not clear from this paper what is the distribution of their nodes and how they are clustered to the boundary. As this paper is detailed in the monograph [49] it seems that they use a uniform grid. Our strategy takes the advantage of the Chebyshev clustering to the boundary.

In our numerical experiments we compute solutions of Mathieu's systems (3.43)–(3.46) using ChC discretizations. For each of the four systems we know from [46] that for every pair of nonnegative indices (i, j) there exists a pair (a_{ij}, q_{ij}) with non-zero functions F_{ij} and G_{ij} such that G_{ij} has exactly i zeros on $(0, \frac{\pi}{2})$ and F_{ij} has exactly j zeros on $(0, \xi_0)$. This is one way how we can index the solutions.

Another option of indexing comes from the fact that Mathieu's systems (3.43)–(3.46) are related to the problem of a vibrating elliptic membrane with fixed boundaries (3.47). Each solution (a, q) gives an eigenmode of (3.47) with the eigenfrequency $\omega = 2\sqrt{q}/h$. The solutions of (3.43) and (3.44) give all even eigenmodes of (3.47). We order the even eigenmodes so that $\omega_1^e \le \omega_2^e \le \cdots$. To each even eigenmode (see, for example, [31, 50]) we can associate an index (k, l), where k is the number of zeros of G on $(0, \pi)$, and l is the number of zeros of F on $(0, \xi_0)$. The eigenmode is then $\psi_e^{k,l}(x, y) = F(\xi)G(\eta)$. In a similar way the solutions of (3.45) and (3.46) give the odd eigenmodes $\psi_o^{k,l}$ of (3.47).

In particular, if F_{ij} and G_{ij} are solutions of one of Mathieu's systems (3.43)–(3.46), then

$$F_{ij}(\xi)G_{ij}(\eta) = \begin{cases} \psi_e^{2i,j}(x,y) & \text{for (3.43),} \\ \psi_e^{2i+1,j}(x,y) & \text{for (3.44),} \\ \psi_o^{2i+2,j}(x,y) & \text{for (3.45),} \\ \psi_o^{2i+1,j}(x,y) & \text{for (3.46).} \end{cases}$$

The choice of the method to solve the algebraic MEP depends on the requested eigenvalue and the required accuracy. It is clear that if we want to compute a higher eigenfrequency very accurately, we need a larger N. Depending on the size of N, we propose to use one of the following methods (see our paper [16])

(a) EIG-Γ: When N is small, we can apply the existing numerical methods (for instance eig in MATLAB) to the eigenvalue problem

$$\Delta_0^{-1}\Delta_2 \mathbf{z} = \mu \mathbf{z}, \tag{3.59}$$

where the matrix $\Gamma_2 := \Delta_0^{-1}\Delta_2$ is of size $N^2 \times N^2$. The obtained eigenvector $\mathbf{z}$ is decomposable, i.e., $\mathbf{z} = \mathbf{x} \otimes \mathbf{y}$, and it is easy to compute $\mathbf{x}$ and $\mathbf{y}$ from $\mathbf{z}$ (see the Appendix A).

(b) EIGS-Γ: When matrix Γ_2 is too large for (a), we can apply the implicit shift-and-invert Arnoldi (available as function eigs in MATLAB) on (3.59). One can see that the matrix Γ_2 is quite sparse. It has N full blocks of size $N \times N$ on its diagonal, whereas all non-diagonal $N \times N$ blocks are diagonal matrices. In many cases, when we need just a small number of eigenvalues, (b) is more efficient than (a) even for a small N.

If N is very large, this approach is not feasible anymore. The first problem is that the L and U factors of the LU decomposition of the matrix $\Gamma_2 - \sigma I$ are virtually full triangular matrices, and we run out of memory.

Although we could try to use another solver instead of the default LU decomposition in eigs, there is another problem when N is large. Namely, as the matrix Γ_2 has size $N^2 \times N^2$, the method builds its search space by vectors of size N^2, which is time and memory consuming.

(c) $JD-W$: When N is too large for (b), we can apply the JD method. An advantage of the JD method is that it works with matrices and vectors of size N. Therefore, the method might be applied when (b) is too expensive.

The results were obtained using MATLAB R2011b running on Intel Core Duo P8700 2.53 GHz processor using 4GB of memory. In this environment, the approach EIGS-Γ works up to $N = 80$, for larger N we have to use $JD - W$. The method EIGS-Γ might be more efficient than EIG-Γ if many eigenvalues are required. MATLAB implementations of the algorithms are available on e-mail request.

Example 3.1 We compare EigElip, which is a MATLAB implementation of our method EIGS-Γ, to the MATLAB function runelip from [48], which was used to compute the eigenfrequencies in [50]. Table 3.6 contains the results for the computation of the n lowest even eigenfrequencies for the ellipse with given α and β. Parameters N_1 and N_2 for EigElip specify the number of points used for the

Table 3.6 Comparison of EigElip and runelip

α	β	n	EigElip			runelip		
			(N_1, N_2)	Time	Error	nrts	Time	Error
2	1	100	(54, 25)	2.5	3e−11	(26,6)	10.3	3e−11
2	1	200	(66, 32)	8.5	2e−11	(38,9)	23.6	5e−11
2	1	300	(80, 36)	23.2	3e−11	(48,11)	37.7	6e−11
2	1	400	(85, 40)	42.0	5e−11	(56,13)	54.0	6e−11
2	1	500	(93, 45)	78.4	3e−11	(63,14)	70.0	3e−01
4	1	100	(68, 24)	3.5	5e−11	(35,5)	12.5	2e−11
4	1	200	(86, 26)	10.2	5e−11	(50,6)	22.1	3e−11
4	1	250	(94, 28)	16.1	3e−11	(56,7)	29.0	3e−06
4	1	300	(100, 30)	24.8	5e−11	(62,8)	36.6	3e−03
8	1	100	(84, 18)	3.1	2e−11	(48,3)	11.1	2e−11
8	1	125	(94, 20)	5.3	2e−11	(55,4)	17.2	3e−05
8	1	150	(100, 20)	6.9	1e−11	(60,4)	19.4	1e−02
cosh(2)	sinh(2)	100	(39, 43)	3.7	2e−11	(24,9)	10.0	1e−11

discretization of Mathieu's systems, which might be different for each of the two equations. The number N_1 is used for the angular equation and N_2 is used for the radial equation. The values are chosen so that the computed eigenfrequencies are correct to at least 10 decimal places. The parameter nrts $= (k_m, l_m)$ in runelip specifies that the method computes all eigenfrequencies of index (k, l) where $k \leq k_m$ and $l \leq l_m$. The values are minimal possible so that all of the n lowest eigenfrequencies are among the computed ones. The computational times, which are given in seconds, show that the new method is considerably faster than runelip for a modest n. For large n runelip can be faster than EigElip (see $\alpha = 2$, $\beta = 1$, and $n = 500$), but also less accurate. The values in the 6th and the 9th column present the maximum absolute error of the computed eigenvalues, where the "exact" eigenvalues to compare with were computed with larger N_1 and N_2. One can see that runelip becomes inaccurate for higher eigenfrequencies, in particular when the ratio α/β is large (see also Example 3.4).

Example 3.2 In this example we use *JD* with harmonic Ritz values, presented in the Appendix A, to compute eigenvalues close to a given target. Depending on the region of interest we do this for several targets. The result of this phase is a set of eigenpairs $((\lambda_k, \mu_k), \mathbf{x}_k \otimes \mathbf{y}_k)$ for $k = 1, \ldots, m$. For each obtained eigenvalue (λ_k, μ_k) we compute its index (i_k, j_k), where i_k and j_k are the number of zeros of $\mathbf{x}_k$ and $\mathbf{y}_k$, respectively. Here we assume that vectors $\mathbf{x}_k$ and $\mathbf{y}_k$ are discrete approximations of continuous curves.

   In the second phase we extend the obtained set by the TRQI. We exploit the following property of eigenvectors of Mathieu's system. Let $\mathbf{x}_1 \otimes \mathbf{y}_1$ and $\mathbf{x}_2 \otimes \mathbf{y}_2$ be approximate eigenvectors belonging to the eigenvalues with indices (i_1, j_1) and (i_2, j_2), respectively. If $j_1 = j_2$ and i_1 is close to i_2, then x_1 and x_2 do not differ much.

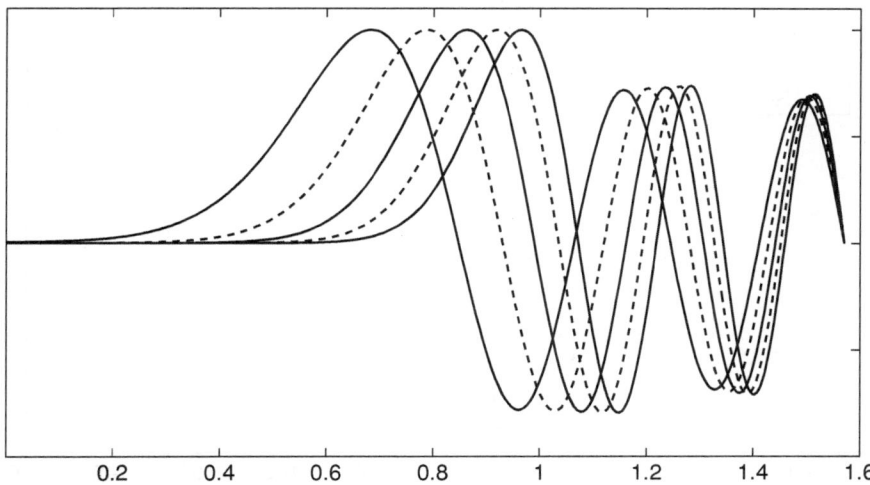

Fig. 3.17 x-part of eigenvector of (3.51) corresponding to the eigenvalue with index $(4, j)$ for $j = 2, \ldots, 6$

The same applies to $\mathbf{y}_1$ and $\mathbf{y}_2$ when $i_1 = i_2$ and j_1 is close to j_2. This is displayed in Fig. 3.17, where the $\mathbf{x}$ part of the eigenvector corresponding to the eigenvalue with index $(4, j)$ is presented for $j = 2, \ldots, 6$. So, for each pair of eigenvectors, such that i_1 is close to i_2 and j_1 is close to j_2, we can apply TRQI with an initial approximation $\mathbf{x}_1 \otimes \mathbf{y}_2$ to compute the eigenpair with the index (i_1, j_2). This simple approach usually converges in a couple of steps.

We take the ChC discretization (3.51) with matrices of size 100×100 and two targets: $(0, 0)$ and $(100, 0)$. For each target we do 200 outer iterations of the *JD* method with harmonic Ritz values. As preconditioners we take $(A_i - \sigma_T B_i - \tau_T C_i)^{-1}$ for $i = 1, 2$, where (σ_T, τ_T) is the current target. We apply 20 steps of the GMRES to solve the correction equations. As a result, we get 44 eigenpairs for the target $(0, 0)$ and 27 eigenpairs for the target $(100, 0)$. From these eigenpairs we compute additional 13 eigenvalues with the TRQI, so that in the end we have approximations for all eigenpairs of (3.43) with indices (i, j), such that $i \leq 13$ and $j \leq 5$.

The computed eigenvalues are presented in Fig. 3.18 with different symbols. The eigenvalues marked by dot and $\times$-mark were computed using the *JD* method with targets $(0, 0)$ and $(100, 0)$, respectively, while the eigenvalues marked by plus were computed using the TRQI. One can see that in a large majority the eigenvalues computed by the *JD* method are indeed the closest ones to the given target.

Example 3.3 Using the method EIGS-Γ we can accurately compute higher eigenmodes than the previously reported in the literature. For example we take $\alpha = 4$ and $\beta = 1$ and compute the lowest 300 even eigenmodes using EigElip with $N_1 = 120$ and $N_2 = 40$. The eigenmodes $\psi_e^{3,8}$ and $\psi_e^{52,3}$ for the eigenfrequencies

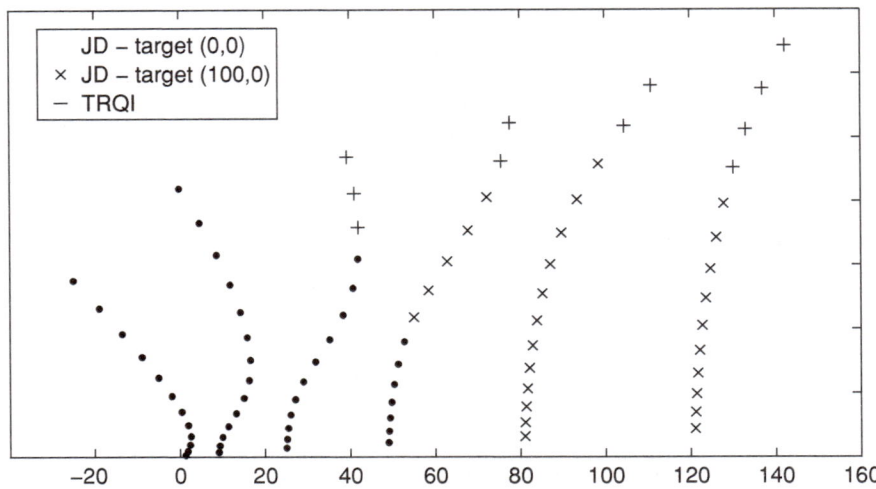

Fig. 3.18 All eigenvalues of (3.43) with indices (i, j) for $i \leq 13$ and $j \leq 5$, computed by the Jacobi-Davidson method using targets $(0, 0)$ and $(100, 0)$, and extended by the TRQI

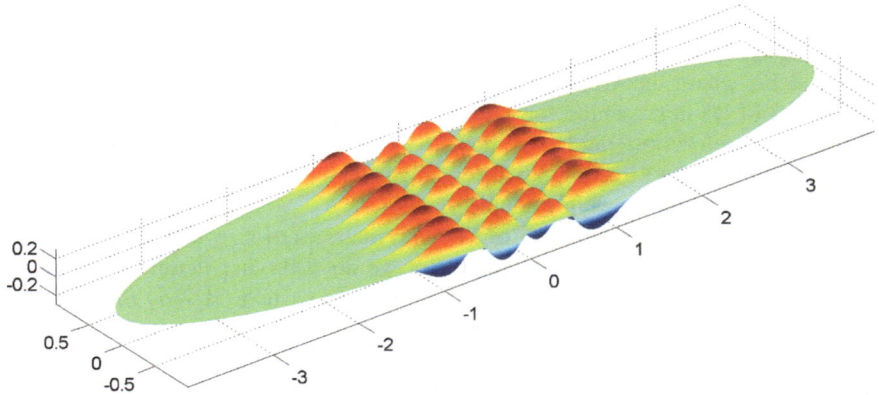

Fig. 3.19 Eigenmode $\psi_e^{3,8}$ for the ellipse with $\alpha = 4$ and $\beta = 1$

$\omega_{298}^e = 24.45490912$ and $\omega_{300}^e = 24.53067377$ are presented in Figs. 3.19 and 3.20, respectively.

It is important to underline that even higher eigenmodes, which require larger matrices, could not be obtained by EIG-Γ and EIGS-Γ methods due to memory limitations, while JD-W is able to compute these eigenmodes up to the required accuracy.

Using $N_1 = N_2 = 500$, the corresponding Γ_2 matrix (that we do not compute explicitly) has dimension $250,000 \times 250,000$. For the target $(1, 7500)$ we computed the even eigenfrequency of the ellipse with $\alpha = 2$ and $\beta = 1$ that is closest to the

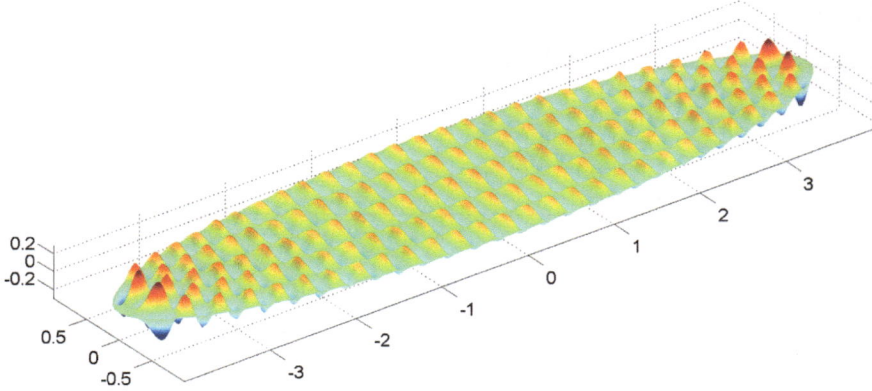

Fig. 3.20 Eigenmode $\psi_e^{52,3}$ for the ellipse with $\alpha = 4$ and $\beta = 1$

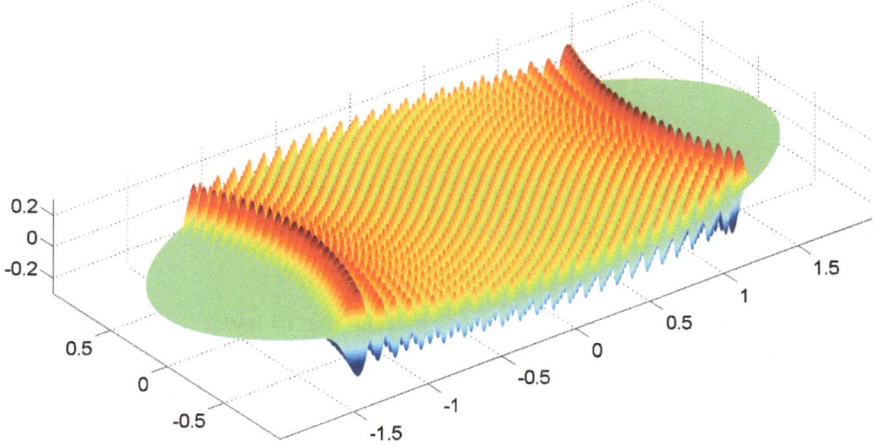

Fig. 3.21 Eigenmode $\psi_e^{41,25}$ for the ellipse with $\alpha = 2$ and $\beta = 1$ that has its eigenfrequency closest to $\omega = 100$

target $\omega_T = 100$. The result is $\omega = 99.97702290$. Its eigenmode $\psi_e^{41,25}$ is presented on Fig. 3.21.

Example 3.4 There are certain applications that require the eigenmodes of an ellipse with a very large ratio α/β (see, for instance, [29]).

In this example we show that such problems can also be solved efficiently via the MEP approach. If we take the ellipse with $\alpha = 1,000$ and $\beta = 1$, and set $N_1 = 200$ and $N_2 = 15$, then `EigElip` returns the 10 lowest even eigenfrequencies in Table 3.7. The 6th lowest even eigenmode is shown in Fig. 3.22.

Table 3.7 The lowest 10 even eigenfrequencies for the ellipse with $\alpha = 1,000$ and $\beta = 1$

n	Eigenfrequency	n	Eigenfrequency
1	1.57129649	6	1.57630126
2	1.57229680	7	1.57730317
3	1.57329744	8	1.57830539
4	1.57429840	9	1.57930793
5	1.57529967	10	1.58031079

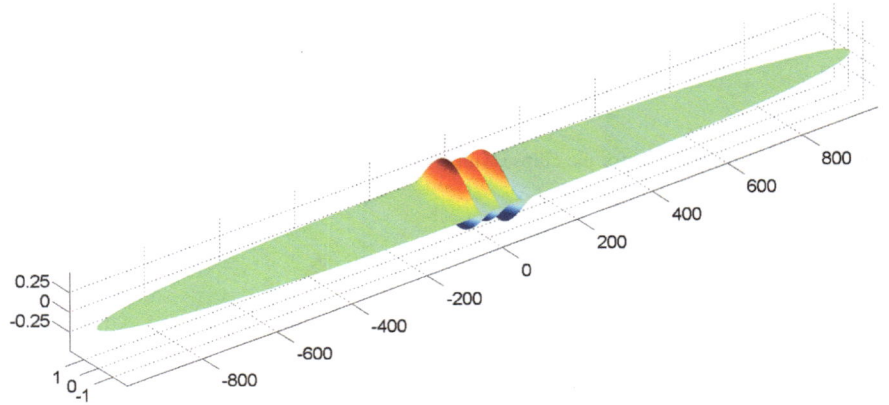

Fig. 3.22 Eigenmode $\psi_e^{6,1}$ for the ellipse with $\alpha = 1,000$ and $\beta = 1$

Let us remark that `runelip` fails to compute the 10 lowest eigenfrequencies. It returns 0, followed by 4 eigenfrequencies smaller than 1. Such results provide confidence in the validity of our approach.

Example 3.5 We compare the accuracy of the computed eigenfrequencies if we discretize the Mathieu's system (3.43) by the ChC discretization (3.51) or by the standard FD scheme (3.58). We take $\alpha = 2$, $\beta = 1$ and compute even eigenfrequencies ω_1^e, ω_{50}^e, and ω_{100}^e. The absolute errors for the ChC and for the FD are collected in Tables 3.8 and 3.9, respectively, where the "exact" eigenvalues were computed by the ChC method using larger N_1 and N_2. It is obvious that in spite of better conditioned matrices involved (symmetric, tridiagonal etc.) finite differences require much larger matrices to obtain accurate results. On the other hand, using the ChC scheme we can compute eigenvalues quite accurately with relatively small matrices.

It is worth noting that the largest Γ_2 matrix corresponding to FD discretization has dimension $1600^2 \times 1600^2$ and the eigenvalue problem can only be solved using *JD* algorithm.

We end up this section with a collection of eigenmodes. Some 2π even radial and angular modes are depicted in Fig. 3.23 and respectively Fig. 3.24. A 3D view of

Table 3.8 Accuracy of the ChC method

N	Error for ω_1^e	Error for ω_{50}^e	Error for ω_{100}^e
(20, 10)	1.8e−10	1.4e−2	2.8e−3
(30, 15)	8.1e−14	5.9e−6	1.4e−4
(40, 20)	9.5e−14	3.0e−10	8.0e−8
(50, 25)	2.1e−13	1.2e−14	1.2e−11

Table 3.9 Accuracy of FD scheme

N	Error for ω_1^e	Error for ω_{50}^e	Error for ω_{100}^e
200	5.9e−3	5.7e−2	3.9e−2
400	3.0e−3	2.6e−2	1.2e−2
800	1.5e−3	1.2e−2	4.5e−3
1,600	7.4e−4	6.0e−3	1.8e−3

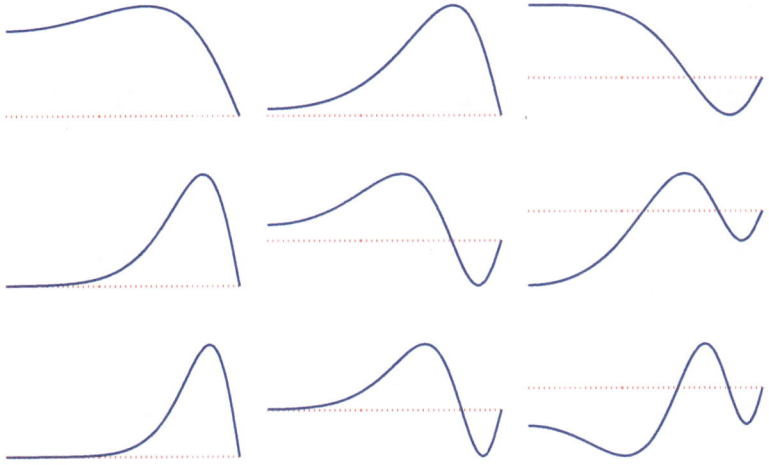

Fig. 3.23 Some 2π even radial modes

some solutions of the problem (3.44) are depicted in Fig. 3.25 and their projections are provided in Fig. 3.26.

3.6 Improvements Induced by *JD* Methods

JD methods proved to be useful as well as fairly accurate in solving large algebraic MEPs associated with the discretizations of Mathieu's system. Actually they provide the same benefits when various GEPs are considered. This section tries to justify this statement.

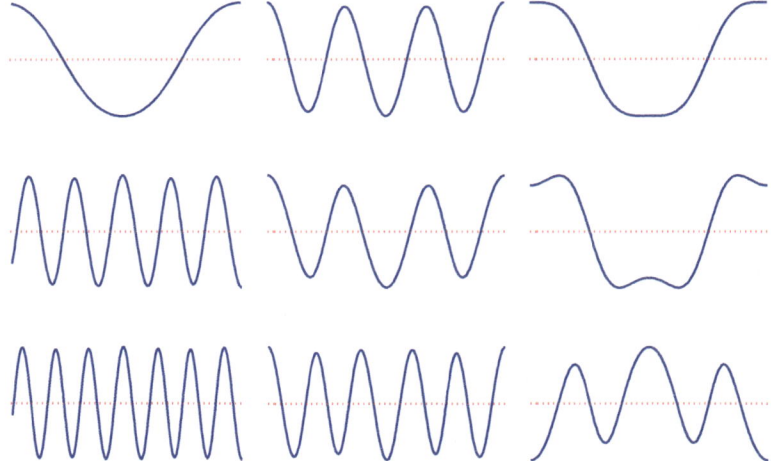

Fig. 3.24 Some 2π even angular modes

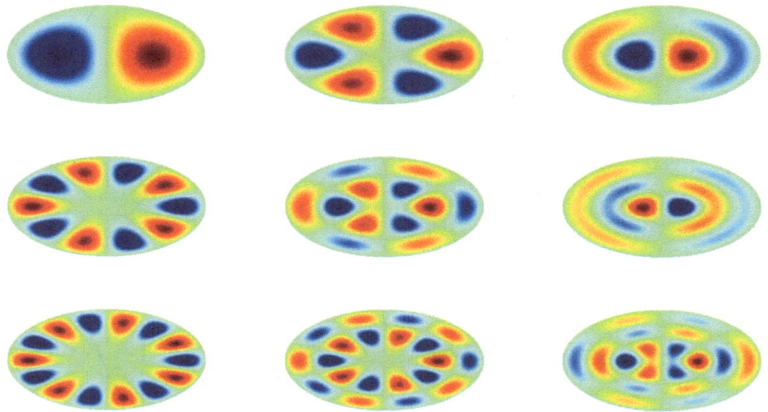

Fig. 3.25 A 3D view of solutions of the problem (3.44)

In a linear hydrodynamic stability analysis one first considers the linearization of the governing equations around a steady-state flow. The perturbation variables are described by a linear system of coupled partial differential equations supplied with a set of boundary conditions. The discretization of this system frequently leads to some *non-Hermitian GEPs* with *singular second order matrix in pencil* of the form

$$A\mathbf{x} = \lambda B\mathbf{x}, \tag{3.60}$$

where the matrices A and B satisfy $rank\, B < rank\, A$. The singularity of B can have two different reasons. First reason can be physical, i.e., in incompressible flows this

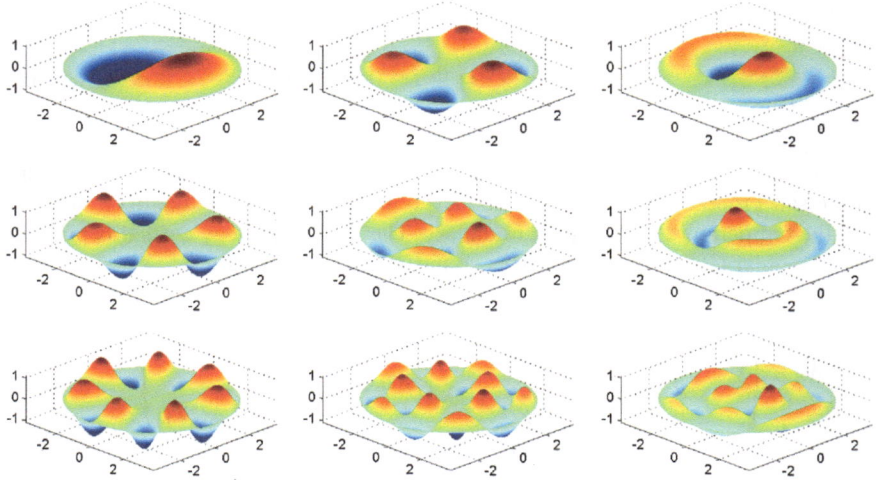

Fig. 3.26 Contour plots of solutions of the problem (3.44)

matrix is associated with the transition terms in governing equations and then, the singularity comes from the fact that continuity equation for such flows does not have a time dependent term. Second, the singularity comes from the numerical method of discretization. We are mainly interested in this second situation.

However, this singularity is responsible for the so called *spurious eigenvalues* or *eigenvalues at infinity*. They complicate the numerics because most iterative methods favor the eigenvalues with the largest modulus, not those with largest real part. There are many examples of such problems but we restrict ourselves to the following three.

Three singular GEPs In the previous sections we have encountered four singular GEPs namely (2.43), (3.12), (3.39), (3.21) and (3.22). Let us write this last problem as

$$A\mathbf{x} = Ra B\mathbf{x}, \tag{3.61}$$

where the matrices A and B are defined by (3.27) and consider it as a test one in working with *JD* methods.

As a second test problem we consider another representative singular hydrodynamic stability problem. In this case we solve an eight order differential system which models *multi-component convection-diffusion in a porous medium*. Defining a^2 to be the wave number and λ the eigenparameter, the non-dimensional linear perturbation equations read [25]

$$\begin{cases} \left(D^2 - a^2\right) W - 2\left(\zeta - z\right) a^2 S - a^2 \Psi^1 - a^2 \Psi^2 = 0, \\ \left(D^2 - a^2\right) S - RW = \lambda S, \\ \left(D^2 - a^2\right) \Psi^1 - R_1 W = \lambda P_1 \Psi^1, \\ \left(D^2 - a^2\right) \Psi^2 - R_2 W = \lambda P_2 \Psi^2, \quad z \in (0, 1), \end{cases} \tag{3.62}$$

supplied with the boundary conditions

$$W = S = \Psi^1 = \Psi^2 = 0, \ z = 0, 1. \tag{3.63}$$

In the system (3.62), W, S, Ψ^1, Ψ^2 are the perturbations of velocity, temperature and two solutes, R and R_β are thermal and solute Rayleigh numbers, respectively, and P_β are salt Prandtl numbers, $\beta = 1, 2$. Eventually ζ signifies a quantity connected with the temperature of the upper boundary. The "$D^{(2)}$" strategy casts this problem into a singular GEP with the matrices A and B defined as follows

$$A := \begin{pmatrix} 4\widetilde{D}^{(2)} - a^2 - 2\,(\zeta - z)\,a^2 I & -a^2 I & -a^2 I \\ -RI & 4\widetilde{D}^{(2)} - a^2 I & Z & Z \\ -R_1 I & Z & 4\widetilde{D}^{(2)} - a^2 I & Z \\ -R_2 I & Z & Z & 4\widetilde{D}^{(2)} - a^2 I \end{pmatrix},$$

$$B := \begin{pmatrix} Z & Z & Z & Z \\ Z & I & Z & Z \\ Z & Z & P_1 I & Z \\ Z & Z & Z & P_2 I \end{pmatrix}. \tag{3.64}$$

As a third test problem we consider the stability of the so called *Hadley flow*. It refers to convection in a layer of porous medium where the basic temperature field varies in the vertical (i.e., z direction) as well as along one of the horizontal directions which is defined as x direction. The non-dimensional perturbation equations are [25]

$$\begin{cases} (D^2 - a^2)\,W + a^2 S = 0, \\ (D^2 - a^2 - i\sigma - ikU\,(z))\,S + ika^{-2}R_H DW - (DT)\,W = 0, \ z \in \left(-\frac{1}{2}, \frac{1}{2}\right), \end{cases} \tag{3.65}$$

supplied with boundary conditions

$$W = S = 0, \ z = \pm\frac{1}{2}. \tag{3.66}$$

In (3.65), $W(z)$ and $S(z)$ are respectively the third component of velocity and temperature field perturbations. The steady-state solutions have the form

$$U\,(z) := R_H z,$$
$$T\,(z) := -R_V z + \tfrac{1}{24} R_H^2 \left(z - 4z^3\right) - R_H x,$$

where R_H and R_V are the horizontal and vertical Rayleigh numbers, respectively, $a^2 := k^2 + m^2$ with k and m being the x and y wave numbers and σ is the eigenparameter. The same strategy casts this problem into a GEP one with the matrices A and B defined as follows

Fig. 3.27 Log of the condition number of the matrices $\widetilde{D}^{(2)} - a^2 I$ (*square line*), $\left(\widetilde{D}^{(2)} - a^2 I\right)^2$ (*diamond line*), $\left(\widetilde{D}^{(2)} - a^2 I\right)^3$ (*hexagram line*) and $\left(\widetilde{D}^{(2)} - a^2 I\right)^4$ (*circle line*) versus N in ascending order; $a = 4.92$

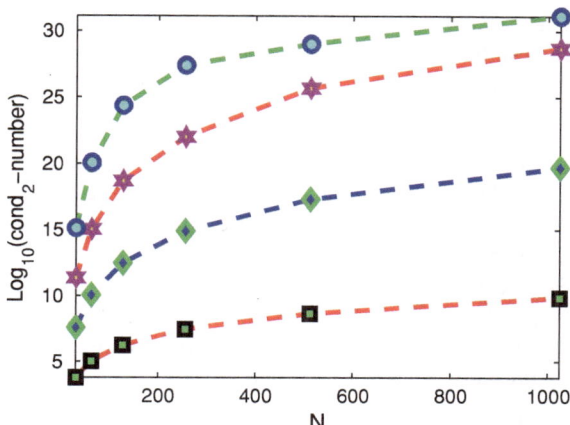

$$A := \begin{pmatrix} 4\widetilde{D}^{(2)} - a^2 I & a^2 I \\ (-\frac{R_H^2}{24} + R_V)I + (\frac{R_H^2}{2})X + (ik\frac{R_H^2}{a^2})\widetilde{D}^{(1)} & 4\widetilde{D}^{(2)} - a^2 I - ik\,\mathbf{diag}(\frac{R_H x}{2}) \end{pmatrix},$$

$$B := \begin{pmatrix} Z & Z \\ Z & I \end{pmatrix},$$

(3.67)

where $X := \mathbf{diag}(\mathbf{x}_{int}^2/4)$, and $\widetilde{D}^{(1)}$ is the first order differentiation matrix. We remark that the problem (3.65, 3.66) can be reduced to a fourth order one in W containing the leading term $\left(D^2 - a^2\right)^2$.

Our factorization is strongly motivated by the conditioning of the matrices A, $\left(4\widetilde{D}^{(2)} - a^2 I\right)$ and its second, third and fourth powers. In Fig. 3.27 the log of the condition numbers of the last four matrices is depicted. It is extremely important to observe that the condition number of $\left(\widetilde{D}^{(2)} - a^2 I\right)$ is of order $O(N^3)$ which means a much better conditioned matrix than the matrix $\widetilde{D}^{(2)}$. We recall that the condition number of the latter matrix is $O(N^6)$ (see Fig. 3.13 in Sect. 3.3). The explanation consists in the fact that the term $\left(-a^2 I\right)$ acts like a *preconditioner* for the differentiation matrix $\widetilde{D}^{(2)}$.

It is the main reason for which we have not applied directly collocation scheme to the problem (3.28) or to its weak formulation (3.29) or even to its minimization formulation (3.31). A completely analogous remark holds for the problems (3.62, 3.63) and (3.65, 3.66).

Numerical results carried out with *JD* algorithm It is well established that, the major drawback of full space methods such as the QR method (for ordinary eigenvalue problems) and the QZ method (for GEPs, see [19–21]) is their computational complexity of $O(n^3)$, where n is the dimension of the problem. Consequently, for $n \gg 5000$, direct computation of the complete spectrum is no longer feasible using the QR or QZ method (see the uppermost curve in Fig. 3.28).

On the other hand, for the applications described in this section (and for many other applications as well), one is not interested in the *complete* spectrum, but in a *few specific* eigenvalues: the eigenvalues closest to zero. In fact the eigenvalue closest to zero is the most important. Before describing the proposed algorithm, first the requirements for an eigensolution method for the applications we aim at are summarized:

1. Given the number k of wanted eigenvalues and a target (e.g., closest to zero), the methods must produce k eigenvalues closest to the target.
2. The method must be scalable with respect to the dimension of the eigenvalue problem.
3. The method must produce *real* approximate values for *real* eigenvalues and *complex* (conjugated pairs of) approximate values for *complex* eigenvalues.
4. The method must not be hampered by eigenvalues at infinity of the eigenvalue problems (3.61), (3.64) and (3.67).

The *JD* method is analyzed in [37], as an iterative method for the computation of a few specific eigenvalues and was reviewed in our previous paper [17].

In this section we will report some representative numerical results obtained in solving the algebraic GEPs defined by Eqs.(3.61), (3.64) and (3.67). All numerical computations reported in [9, 15] were carried out using the MATLAB code `eig`, which is an implementation of the QZ method. Alternatively, we use now the $JDQZ$ implementation from [36], the real variant $rJDQZ$ implementation described in [45], and then compare the results with those obtained using the MATLAB code `eig` for the QZ method (see also Table 3 in [15]) and the MATLAB code `eigs` for the Arnoldi method. As far as we know and as the MATLAB help shows, the `eigs` implementation relies on the IRAM code of ARPACK library. All computations were carried out using MATLAB 2010a on a HP xw8400 Workstation with clock speed of 3.2 GHz.

The computed values of the first three eigenvalues, i.e., $R_1, R_2, R_3, \lambda_1, \lambda_2, \lambda_3$ corresponding to problems (3.61) and (3.64) respectively are displayed in Table 3.10. The eigenvalues $\sigma_1, \sigma_2, \sigma_3$ corresponding to problem (3.67) are reported in Table 3.11. The CPU times reported in these tables, as well as in Fig. 3.28, measure the time required for the computation of the first six eigenvalues except QZ method which computes the whole spectrum.

They are obtained for the following sets of fixed physical parameters: $a^2 = 21.344$, $\zeta = 0.14286$, $R = 228.009$, $R_1 = -291.066$, $R_2 = 261.066$, $P_1 = 4.5454$, $P_2 = 4.7619$, in case of problem (3.64), $\varepsilon = 0.03$ and $a = 4.92$ in case of problem (3.61), and $k = 0$, $m = 10$, $R_H = 114.2$, $R_V = 100$, in case of problem (3.67).

It is fairly clear from both tables that numerical approximations of the first three leading eigenvalues computed by both $JDQZ$, QZ and Arnoldi are almost indistinguishable.

The CPU times required by the all four methods are depicted in Fig. 3.28. It is important to observe that $rJDQZ$ computed only the first two eigenvalues for spectral parameter N larger than 1,000.

With respect to the memory usage, we have to mention that for all three methods Arnoldi, JD and rJD this is of order $O(kn)$, k being the maximal dimension of the

Fig. 3.28 CPU time versus
$n = (N - 2) * 4$ for problem
(3.64). *Star line* corresponds to
rJDQZ, *circle line* to Arnoldi,
diamond line to *JDQZ* and
square line to *QZ*

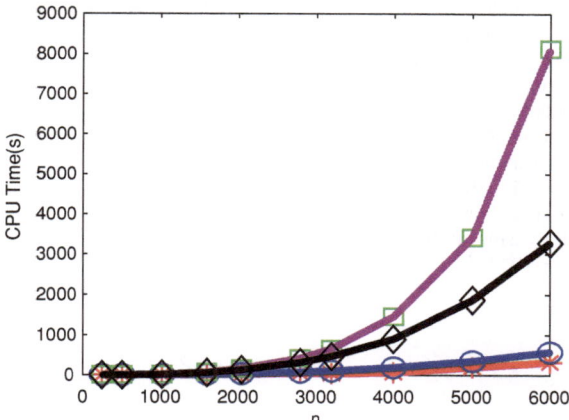

search space, and $O(n^\alpha)$ with $1 < \alpha \leq 3$ for the system solves, where for sparse systems $\alpha < 2$ may be expected. For $\alpha > 2$ costs increase rapidly, and JD/rJD would be only alternative since inexact solves could be used (e.g. GMRES, of course of costs $O(mn)$ where m is the number of GMRES iterations).

However, the actual performance of *JD* relies on many factors, including the size of the search spaces, construction of projection spaces, shift selection (Ritz, Petrov, harmonic) and the accuracy of correction equation solves.

The real variant of $JDQZ$, $rJDQZ$ does produce real eigenvalues because it uses real search spaces.

Both Arnoldi and the *JD* based methods did not show any difficulties with spurious eigenvalues.

We also have to emphasize that, as far as we know, we have used the largest spectral order N in such eigenvalue computations. Comparatively, in [25, 39] values of N attain 50, in [2] the authors work with N of order 100, in [43] the authors consider that $N = 602$ is fine enough to predict correctly the leading eigenvalues of the problem and in [30] this order attains 1,000. In older papers the authors simply work with N much less than 100.

Some more comments on the convergence and round-off errors in $rJDQZ$ and Arnoldi method when singular GEPs are solved are now in order. We define first, the error of a method as the absolute value of the difference between the computed eigenvalue and the converged one, i.e., obtained with the largest resolution.

Then, we observe that the pseudospectra of our pencils (A, B) can be unbounded, i.e., they extend to the whole complex plane for sufficiently large perturbations (see [44]). Thus, to get some information we cut up from the whole spectrum the region surrounding the first two eigenvalues and observe that our computation is *backward stable* in the sense that the line of level which equals the error encloses the computed eigenvalue. Using a resolution of 256 which implies a matrix of order $1 \times 10^{+03}$ the pseudospectrum is depicted in Fig. 3.29. As the diameter of the enclosed area

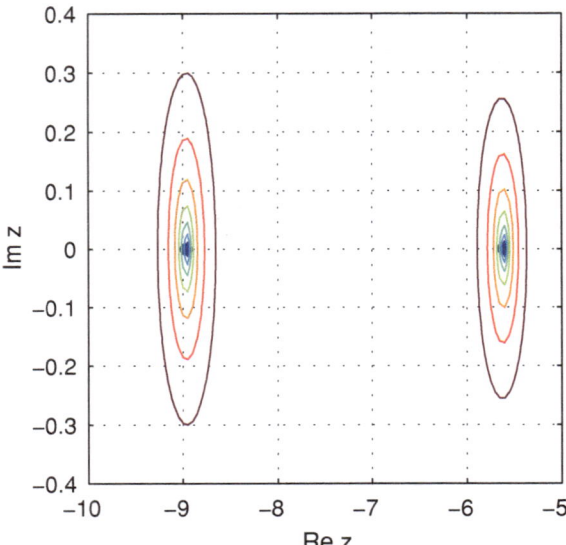

Fig. 3.29 The part of pseudospectrum surrounding the first two eigenvalues of problem (3.64) obtained using $rJDQZ$. The largest contour line corresponds to an error of order 1×10^{-01} and the nested contours correspond to errors decreasing successively with $1 \times 10^{-0.02}$

provides a hint of the largest possible relative error on the respective eigenvalue, we can infer a more sensitive second eigenvalue than the first one.

The errors of the first two computed eigenvalues, by $rJDQZ$ and respectively Arnoldi, versus $1/N$, as is usual in spectral methods, are depicted in Fig. 3.30. For resolutions larger than 1,500 this picture shows that the computation of the second eigenvalue becomes unstable in both methods. It is in accordance with the conclusion provided by the pseudospectrum. In order to observe the order of convergence of both methods we depicted Fig. 3.31. The $rJDQZ$ method for resolutions inferior to 1700 has order of convergence that could attains unity, i.e., Err is of order $(1/N)^p$, $0 < p \leq 1$. These resolutions are satisfactory in hydrodynamic stability problems. Arnoldi method behaves better. It has a comparable order of convergence but looks more accurate. The large, but still acceptable error of $rJDQZ$ can be explained by the fact that the convergence tolerance was set to 1×10^{-6}. Higher accuracy can be obtained for lower tolerances, at the cost of additional JD iterations. With respect to Arnoldi method we have to observe that this method confirms its performances established in [42] also for singular GEPs.

Consequently, in this section we have shown that the JD and Arnoldi methods can be successfully used to compute the smallest eigenvalues of large-scale matrix pencils arising in linear hydrodynamic stability problems.

Both methods are not hampered by eigenvalues at infinity. Since these methods are applicable to large-scale problems, they are preferred over the QZ method, which is only applicable to moderate size eigenvalue problems. The variant of the JD method that uses only real search spaces, $rJDQZ$, takes advantage of using such search spaces resulting in real approximations of real eigenvalues. Consequently, it is the

Fig. 3.30 Semilog plot of
error versus N for the first two
eigenvalues computed with
$rJDQZ$ respectively Arnoldi

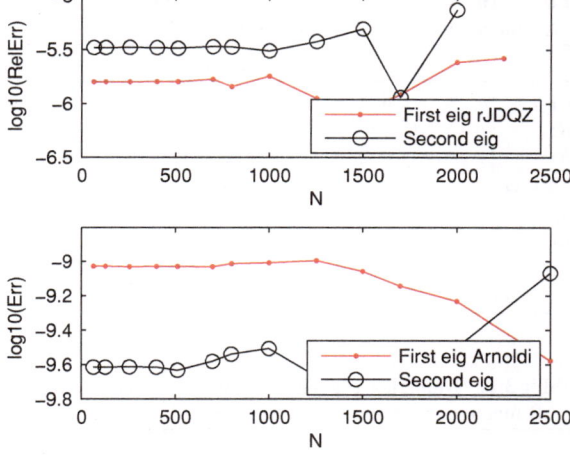

Fig. 3.31 Loglog plot of
error versus $1/N$ for the first
eigenvalues computed with
$rJDQZ$ respectively Arnoldi

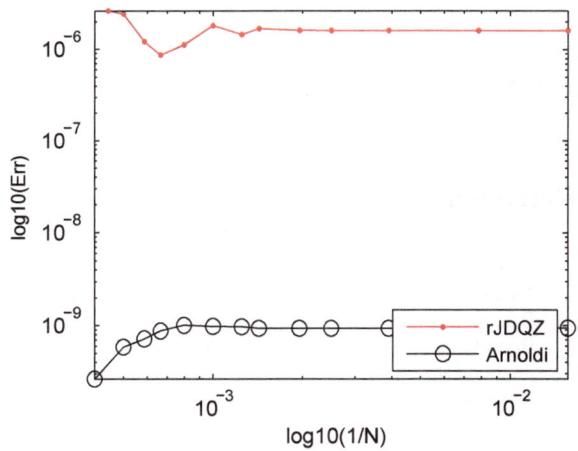

fastest method for solving such large and singular GEPs. Moreover, this method is
completely independent of the physical formulation of the problem at hand.

Eventually, we want to underline that the first eigenvalue of the multi-component
convection-diffusion problem reported in Table 3.10, i.e., $\lambda_1 = -5.60913$, is confirmed up to the last digit in [25]. In the same way the first eigenvalue of the Hadley
flow reported in Table 3.11, i.e., $\sigma_1 = -0.293433$, is confirmed up to the fifth
decimal digit in the above quoted paper. In [25] the authors obtained these values
by a Galerkin type method based on the Legendre polynomials. They argue that the
method is much more efficient than ChT in solving both problems.

Table 3.10 Numerical evaluations of the first three eigenvalues using $JDQZ$, $rJDQZ$ and MAT-LAB functions eig and eigs, for $N = 512$

	$JDQZ$	$rJDQZ$	Eig (QZ)	Eigs (Arnoldi $A^{-1}B$)
R_1	25.8365	25.836514	25.836514	25.836514
R_2	134.387	134.386749	134.386750	134.386749
R_3	412.323	412.323554	412.323571	412.323571
$CPU(s)$	48.7	9.36	57.15	9.75
λ_1	−5.60913	−5.609129	−5.609129	−5.609129
λ_2	−8.96417	−8.964170	−8.964170	−8.964170
λ_3	−11.1226	−11.122552	−11.122552	−11.122552
$CPU(s)$	133.0	16.14	157.12	30.70

Table 3.11 Numerical evaluations of the first three eigenvalues using $JDQZ$, $rJDQZ$ and MAT-LAB functions eig and eigs, for $N = 256$ ($n = 508$)

	JDQZ	rJDQZ	Eig QZ	Eigs (Arnoldi $A^{-1}B$)
σ_1	−0.293433	−0.293432	−0.293432	−0.293432
σ_2	−0.692455	−0.692454	−0.692454	−0.692454
σ_3	−329.278	−329.2778	−311.5122	−329.2778
$CPU(s)$	3.21	1.90	1.38	1.31

References

1. Babuska, I., Osborn, J.: Eigenvalue problems. In: Ciarlet, P.G., Lions, J.L. (eds.) Finite Element Methods (Part 1), Volume II of Handbook of Numerical Analysis, pp. 645–785. Elsevier Science B.V, Amsterdam (1991)
2. Boomkamp, P.A.M., Boersma, B.J., Miesen, R.H.M., Beijnon, G.V.A.: A chebyshev collocation method for solving two-phase flow stability problems. J. Comput. Phys. **132**, 191–200 (1997)
3. Boyd, J.P.: Traps and snares in eigenvalue calculations with application to pseudospectral computations of ocean tides in a Basin bounded by meridians. J. Comput. Phys. **126**, 11–20 (1996). Corigendum **136**, 227–228 (1997)
4. Boyd, J.P., Rangan, C., Bucksbaum, H.: Pseudospectral methods on a semi-infinite interval with application to the hydrogen atom: a comparison of the mapped fourier-sine method with laguerre series and rational chebyshev expansions. J. Comput. Phys. **188**, 56–74 (2003)
5. Boyce, W.E., Di Prima, R.C.: Elementary Differential Equations and Boundary Value Problems, 9th edn. Wiley, India (2009)
6. Chanane, B.: Accurate solutions of fourth order sturm-liouville problems. J. Comput. Appl. Math. **234**, 3064–3071 (2010)
7. Coman, C.D., Haughton, D.M.: On some approximate methods for the tensile instabilities of thin annular plates. J. Eng. Math. **56**, 79–99 (2006)
8. Dongarra, J.J., Straughan, B., Walker, D.W.: Chebyshev tau-qz algorithm for calculating spectra of hydrodynamic stability problems. Appl. Numer. Math. **22**, 399–434 (1996)
9. Dragomirescu, I.F., Gheorghiu, C.I.: Analytical and numerical solutions to an electrohydrodynamic stability problem. Appl. Math. Comput. (2010). doi:10.1016/j.amc.2010.05.028
10. Fichera, G.: Numerical and Quantitative Analysis. Pitman Press, London (1978)
11. Finch, S.R.: Mathieu eigenvalues. http://www.algo.inria.fr/csolve/mthu.pdf (2008). Accessed 15 Aug 2012

12. Fornberg, B.: A Practical Guide to Pseudospectral Mathods. Cambridge University Press, Cambridge (1998)
13. Funaro, D., Heinrichs, W.: Some results about the pseudospectral approximation of one-dimensional fourth-order problems. Numer. Math. **58**, 399–418 (1990)
14. Georgescu, A., Pasca, D., Gradinaru, S., Gavrilescu, M.: Bifurcation manifolds in multipara-metric linear stability of continua. ZAMM **73**, T767–T768 (1993)
15. Gheorghiu, C.I., Dragomirescu, I.F.: Spectral methods in linear stability. application to thermal convection with variable gravity field. Appl. Numer. Math. **59**, 1290–1302 (2009)
16. Gheorghiu, C.I., Hochstenbach, M.E., PLestenjak, B., Rommes. J.: Spectral collocation solutions to multiparameter Mathieu's system. Appl. Math. Comput. **218**, 11990–12000 (2012)
17. Gheorghiu, C.I., Rommes, J.: Application of the jacobi-davidson method to accurate analysis of singular linear hydrodynamoc stability problems. Int. J. Numer. Meth. Fluids. **71**, 358–369 (2013)
18. Giannakis, D., Fischer, P.F., Rosner, R.: A spectral galerkin method for the coupled orr-sommerfeld and induction equations for free-surface mhd. J. Comput. Phys. **228**, 1188–1233 (2009)
19. Golub, G.H., Wilkinson, J.H.: Ill-conditioned eigensystems and the computation of the jordan canonical form. SIAM Rev. **18**, 578–619 (1976)
20. Golub, G.H., Van Loan, C.F.: Matrix Computations, 3rd edn. The Johns Hopkins University Press, Baltimore (1996)
21. Golub, G.H., van der Vorst, H.A.: Eigenvalue computation in the 20th century. J. Comput. Appl. Math. **123**, 35–65 (2000)
22. Greenberg, L., Marletta, M.: Algorithm 775: the code sleuth for solving fourth-order sturm-liouville problems. ACM T. Math. Software. **23**, 453–493 (1997)
23. Griffel, D.H.: Applied Functional Analysis, 2nd edn. Dover Publications Inc, Mineola (1985)
24. Guo, B-y, Wang, Z-q, Wan, Z-s, Chu, D.: Second order jacobi approximation with applications to fourth-order differential equations. Appl. Numer. Math. **55**, 480–502 (2005)
25. Hill, A.A., Straughan, B.: A legendre spectral element method for eigenvalues in hydrodynamic stability. J. Comput. Appl. Math. **193**, 363–381 (2006)
26. Hochstenbach, M.E., Plestenjak, B.: Backward error, condition numbers, and pseudospectra for the multiparameter eigenvalue problem. Linear Algebra Appl. **375**, 63–81 (2003)
27. Hoepffner, J.: Implementation of boundary conditions. http://www.lmm.jussieu.fr/hoepffner/boundarycondition.pdf (2010). Accessed 25 Aug 2012
28. Huang, W., Sloan, D.M.: The pseudospectral method for third-order differential equations. SIAM J. Numer. Anal. **29**, 1626–1647 (1992)
29. Igbokoyi, A.O., Tiab, D.: New method of well test analysis in naturally fractured reservoirs based on elliptical flow. J. Can. Pet. Tehnol. **49**, 1–15 (2010)
30. Melenk, J.M., Kirchner, N.P., Schwab, C.: Spectral galerkin discretization for hydrodynamic stability problems. Computing **65**, 97–118 (2000)
31. Neves, A.G.M.: Eigenmodes and eigenfrequencies of vibrating elliptic membranes: a klein oscillation theorem and numerical calculations. Commun. Pure Appl. Anal. **9**, 611–624 (2004)
32. Quarteroni, A., Saleri, F.: Scientific computing with MATLAB and Octave, 2nd edn. Springer, Berlin (2006)
33. Quarteroni, A., Sacco, R., Saleri, F.: Numerical Mathematics, 2nd edn. Springer, Berlin (2007)
34. Ruby, L.: Applications of the mathieu equation. Am. J. Phys. **64**, 39–44 (1996)
35. Sleeman, B.D.: Multiparameter spectral theory and separation of variables. J. Phys. A: Math. Theor. **41**, 1–20 (2008)
36. Sleijpen, J.L.: http://www.math.uu.nl/people/sleijpen/JD_software/JDQZ.html. Accessed 20 Feb 2011
37. Sleijpen, J.L., van der Vorst, H.A.A.: A jacobi-davidson iteration method for linear eigenvalue problems. SIAM J. Matrix Anal. A. **17**, 401–425 (1996)
38. Straughan, B.: The Energy Method, Stability, and Nonlinear Convection. Springer, New York (1992)

39. Straughan, B., Walker, D.W.: Two very accurate and efficient methods for computing eigenvalues and eigenfunctions in porous convection problems. J. Comput. Phys. **127**, 128–141 (1996)
40. Stewart, G.W., Sun, J.-G.: Matrix Perturbation Theory. Academic Press, New York (1990)
41. Trif, D.: Operatorial tau method for higher order differential problems. Br. J. Math. Comput. Sci. **3**, 772–793 (2013)
42. Valdettaro, L., Rieutord, M., Braconnier, T., Fraysse, V.: Convergence and round-off errors in a two-dimensional eigenvalue problem using spectral methods and arnoldi-chebyshev algorithm. J. Comput. Appl. Math. **205**, 382–393 (2007)
43. Valério, J.V., Carvalho, M.S., Tomei, C.: Filtering the eigenvalues at infinity from the linear stability analysis of incompressible flows. J. Comput. Phys. **227**, 229–243 (2007)
44. van Dorsslaer, J.L.M.: Pseudospectra for matrix pencils and stability of equilibria. BIT **37**, 833–845 (1997)
45. van Noorden, T.L., Rommes, J.: Computing a partial generalized real schur form using the jacobi-davidson method. Numer Linear Algebr. **14**, 197–215 (2007)
46. Volkmer, H.: Multiparameter Problems and Expansion Theorems. Lecture Notes in Mathematics, vol. 1356. Springer, New York (1988)
47. Weideman, J.A.C., Reddy, S.C.: A matlab differentiation matrix suite. ACM Trans. Math. Software **26**, 465–519 (2000)
48. Wilson, H.B.: Vibration modes of an elliptic membrane. Available from http://www.mathworks.com/matlabcentral/fileexchange. MATLAB File Exchange, The MathWorks, Natick (2004)
49. Wilson, H.B., Turcotte, L.S., Halpern, D.: Advanced Mathematics and Mechanics Applications Using MATLAB, 3rd edn. Chapman and Hall/CRC, Boca Raton (2003)
50. Wilson, H.B., Scharstein, R.W.: Computing elliptic membrane high frequencies by mathieu and galerkin methods. J. Eng. Math. **57**, 41–55 (2007)

Chapter 4
The Laguerre Collocation Method

Abstract The chapter introduces first the functional framework corresponding to the spectral collocation method based on Laguerre functions. The main advantage of these functions is the fact that they decrease smoothly to zero at infinity along with their derivatives. We speculate this behavior in imposing boundary conditions at large distances. On the half-line we solve high order eigenvalue problems, linear as well as some genuinely nonlinear third and fourth order boundary value problems. The applications come from fluid mechanics, i.e., Blasius, Falkner-Skan, density profile equation, Ekman boundary layer etc. and foundation engineering. Consequently, we avoid the empiric domain truncation coupled with various numerical technique (mainly shooting) as a strategy to solve such problems. Some second order eigenvalue problems along with singularly perturbed boundary value problems are also considered. A special attention is payed to the influence of the scaling parameter (which maps the half-line into itself) on the repartition of the Laguerre nodes. We manually tune this geometrical parameter in order resolve narrow regions with high variations of solutions, i.e., the so called boundary or interior layers. Consequently, no domain decomposition, domain truncation and shooting have been used in our numerical experiments. Based on the pseudospectra of two GEPs we comment on limitations of the linear hydrodynamic stability analysis. We also observe that the non-normality of a spectral method depends on the discretization (method itself) and at the same time on the bases of functions (polynomials) used.

Keywords Density profile equation · Ekman boundary layer · Falkner Skan problem · Laguerre collocation · Movement of a pile · Physical space differentiation

> *The efforts to compute solutions of two-point boundary value problems on infinite intervals are in finding ways to approximate a stable manifold, i.e., a nonincreasing solution with a limit, and part of this is done by prescribing boundary conditions which put the solution on that manifold.*
> *U M Ascher, R M M Mattheij, R D Russel, Numerical Solution of Boundary Value Problems for Ordinary Differential Equations, 1988*

C.-I. Gheorghiu, *Spectral Methods for Non-Standard Eigenvalue Problems*,   85
SpringerBriefs in Mathematics, DOI: 10.1007/978-3-319-06230-3_4,
© The Author(s) 2014

4.1 LC Solutions to a Third Order Linear Boundary Value Problem on the Half-Line

We introduce the method with the representative problem

$$u_{xxx} - a_1 u_{xx} - a_2 u_x + a_3 u = f, \ x \in (0, \infty), \ u(0) = u'(0) = 0, \ \lim_{x \to \infty} u(x) = 0,$$
(4.1)

where the coefficients $a_1(x)$, $a_2(x)$ and $a_3(x)$, $x \in (0, \infty)$ are smooth enough in order to guarantee a correct application of the *strong collocation technique*.

The choice of basis functions on which test and trial spaces are spanned strongly depends on special properties of the problem at hand, as for instance the asymptotic behavior of the solution when the independent variable tends to infinity. The functional framework for collocation method is that introduced in Sect. 1.2.

Therefore we rely on interpolation operators which are built on the weighted Laguerre polynomials, i.e., on functions of the form

$$e^{-x/2} L_N(x),$$

$L_N(x)$ being the *classical Laguerre polynomial* of order N.

It is important to observe that, the interval $[0, \infty)$ can be mapped to itself by change of variables $\eta = b\widetilde{\eta}$, where b is any positive real number. This will be called the *scaling factor*. Laguerre method therefore contains a free-parameter. It means that the Laguerre interpolating process is exact for functions of the form

$$e^{-bx/2} p(x)$$
(4.2)

where $p(x)$ is any polynomial of degree $N - 1$ or less.

The general Laguerre interpolation process in weighted Sobolev spaces was recently systematically analyzed in the monograph [36]. However, the authors do not exploit the freedom offered by the above free parameter in LC. Instead, in our computations this parameter will play a key role.

However, the best interpolation result remains that from [3] which reads

$$\|I_N u - u\|_{\omega_0} \leq C N^{(1-m)/2} \|u\|_{H^m_{\omega_\tau}}, m \geq 1,$$
(4.3)

where I_N is the usual interpolation operator, the weights ω are defined as $\omega_0 := e^{-x}$, $\omega_\tau := e^{-(1-\tau)x}$, $0 < \tau < 1$ and the norms have been introduced with (1.21).

As our numerical experiments showed that the high order Laguerre differentiation matrices are to some extent polluted by round off errors for $N > 100$ (see also Fig. 4.3) this result seems to be not very encouraging. Consequently, we try to examine numerically the convergence behavior of some typical functions. Thus, we consider the following four examples of exact solutions to (4.1) with $a_1 = a_2 = a_3 = 1$.

Fig. 4.1 The quasi spectral accuracy in approximating the solution (4.4) which decays at infinity without oscillations for various scaling factors b. For each and every b, one curve corresponds to inf norm and another to L_2 norm

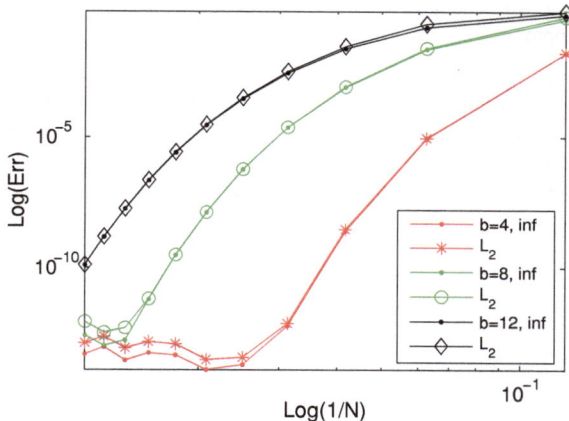

Example 4.1 Exponential decay without oscillations at infinity

$$u(x) := x^2 e^{-x}, \ x \in (0, \infty);$$ (4.4)

Example 4.2 Exponential decay with oscillations at infinity

$$u(x) := e^{-x} \sin^2 kx, \ x \in (0, \infty);$$ (4.5)

Example 4.3 Algebraic decay without oscillations at infinity

$$u(x) := \frac{x^2}{(1+x)^h}, h > 1, \ x \in (0, \infty);$$ (4.6)

Example 4.4 Algebraic decay with oscillations at infinity

$$u(x) := \frac{\sin^2 kx}{(1+x)^h}, h > 1, \ x \in (0, \infty),$$ (4.7)

for some natural k.

These functions have typical decay properties. The errors corresponding to the first case (inf and L_2 norms respectively) are reported in Fig. 4.1 when the scaling factor b as well as the cut off parameter N are simultaneously varied. It is fairly clear that the accuracy depends on both parameters. Whenever this factor is properly chosen, even for moderate values of N, the method produces results with spectral accuracy.

The presence of an oscillation factor deteriorates considerable the approximation and requires a larger interval of approximation. The situation corresponding to the second example is summarized in Fig. 4.2.

Two important conclusions can be formulated. First, as we envisage boundary layer type problems we expect an exponential decay without oscillations of solutions.

Fig. 4.2 The accuracy in
approximating the solution
(4.5) which decays at infinity
with oscillations for various
scaling factors b. For each and
every b, one curve corresponds
to inf norm and another to L_2
norm

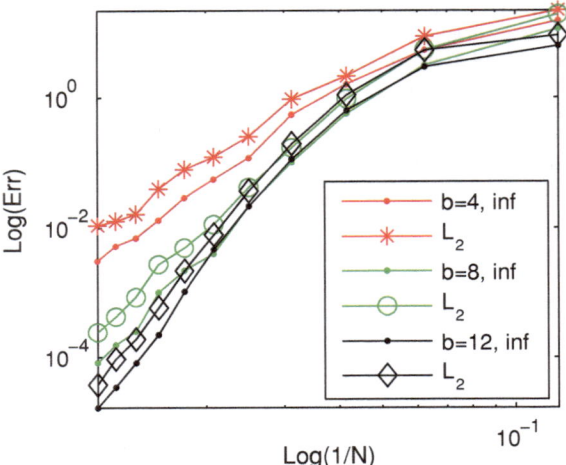

Fig. 4.2 The accuracy in approximating the solution (4.5) which decays at infinity with oscillations for various scaling factors b. For each and every b, one curve corresponds to inf norm and another to L_2 norm

Thus, the spectral accuracy could be altered only by possible nonlinearities of the problem. Second, for oscillatory decaying solutions, *mapping techniques* or even *truncation domain methods* can be a serious concurrent for Laguerre collocation. In [34], and partially in [35], the authors study the accuracy of Laguerre based Galerkin method applied to a self adjoint second order elliptic problem with similar conclusions.

4.2 The Falkner-Skan Problem

The Prandtl's boundary layer equations for flow past a flat plate in a non-uniform external flow (see for instance the well known monographs [26, 32, 33]) are reduced to the third order nonlinear differential equation

$$f''' + ff'' + \beta \left(1 - f'^2\right) = 0, \eta \in (0, \infty), \tag{4.8}$$

where β is a real parameter, the so called *Hartree parameter*, and f' stands for the velocity inside the boundary layer. This is known as the *Falkner-Skan equation*. In fact β signifies the dimensionless pressure gradient parameter and $\beta := \frac{2m}{m+1}$ where m comes from the expression of the free stream velocity $U_\infty(x) := Cx^m$, C being a real constant.

The viscous boundary conditions reduce themselves to the following three

$$f(0) = f'(0) = 0, \quad f' \to 1, \eta \to \infty. \tag{4.9}$$

Whenever there exists heat transfer between the wall and viscous incompressible fluid moving along, the energy equation reduces to the nonlinear second order equation

$$\vartheta'' + \Pr f \vartheta' = 0, \eta \in (0, \infty), \tag{4.10}$$

where ϑ stands for non-dimensional temperature and $\Pr$ for *Prandtl number*. Equation (4.10) is supplied with the following two boundary conditions

$$\vartheta\,(0) = 1, \vartheta \to 0, \quad \eta \to \infty. \tag{4.11}$$

As soon as the problem (4.8, 4.9) is solved, the problem (4.10, 4.11) becomes a linear one and is immediately solvable. The left hand side part of (4.8) is also present in the boundary layer equations of two-dimensional laminar and turbulent flows.

In order to homogenize the boundary condition at infinity in (4.9), we introduce the new unknown $F(\eta)$ by

$$F(\eta) := f(\eta) + 1 - \eta - e^{-\eta}. \tag{4.12}$$

In this new variable Eq. (4.8) and the boundary conditions (4.9) read

$$\left\{ \begin{array}{c} F''' + \left[FF'' + e^{-\eta}\left(F + F''\right) + (\eta - 1)F'' \right] \\ -\beta\left[F'^2 + 2\left(1 - e^{-\eta}\right)F' \right] = g(\eta), \;\; \eta \in (0, \infty), \\ F\,(0) = F'\,(0) = 0, \;\; F' \to 0, \; \eta \to \infty, \end{array} \right. \tag{4.13}$$

where the r.h.s. term $g(\eta)$, becomes

$$g(\eta) = e^{-\eta}\left[\left(2 - \eta - e^{-\eta}\right) - \beta\left(2 - e^{-\eta}\right) \right]. \tag{4.14}$$

4.3 The Laguerre Differentiation Matrices

In order to solve the two-point boundary value problem (4.13) by LC method we represent $F(\eta)$ by the interpolate $p_{N-1}(\eta)$ defined by

$$p_{N-1}(\eta) := \sum_{j=1}^{N} \frac{e^{-\eta/2}}{e^{-\eta_j/2}} \Psi_j(\eta) F_j, \tag{4.15}$$

where $\Psi_j(\eta)$ are the Lagrangian cardinal polynomials

$$\Psi_j(\eta) := \prod_{m=1,\ m \neq j}^{N} \left(\frac{\eta - \eta_m}{\eta_j - \eta_m} \right), j = 1, 2, ..., N. \tag{4.16}$$

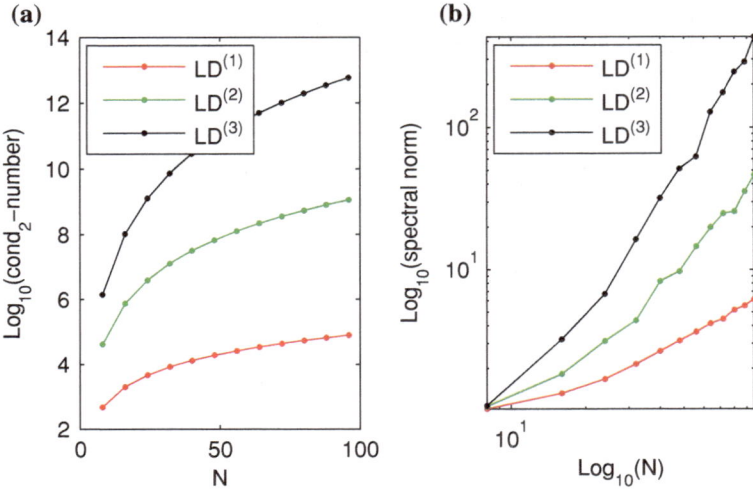

Fig. 4.3 The conditioning of the first three Laguerre differentiation matrices when the scaling factor $b = 2$; **a** the condition number versus N, and **b** the spectral radius versus N

The interpolating nodes η_j, $j = 2, ..., N$ are the roots of Laguerre polynomial of degree $N - 1$, indexed in increasing order of magnitude. We add a node $\eta_1 = 0$ in order to facilitate the incorporation of boundary conditions. Therefore we introduce the nodal unknown values $F_j = F(\eta_j)$, $j = 1, ..., N$.

In order to write down the LC equations for (4.13) we use the Laguerre differentiation matrices from the seminal paper [41] (see also [10]). Let us denote these matrices by $LD^{(n)}$, where $n = 1, 2, 3$ is the differentiation order.

Unfortunately, these matrices are fully populated, non-symmetric and quite bad conditioned. In Fig. 4.3 we report the dependence of their conditioning and of their spectral radius on N.

The Henrici number for the first, second and third order Laguerre differentiation matrices equals respectively the following numerical values: 1.1199, 1.1838 and 1.1882. These values point to a high non-normality. Our numerical experiments have revealed that this scalar measure of non-normality is practically independent of N and the scaling factor.

Thus, at least for spectral collocation method an important conclusion comes out. The non-normality depends on the method (discretization) and for a specified method it depends on the choice of bases of functions (polynomials) involved. The fourth column of Table 3.1 shows a non-normality of approximatively fourth time smaller of ChC differentiation matrices than the values displayed above. The differentiation matrices based on the functions (3.15) are into an intermediate situation.

However, a simple comparison with our results reported in Sects. 3.3 and 3.6 (see Fig. 3.13 and respectively Fig. 3.27) shows better conditioned Laguerre differentiation matrices than their Chebyshev counterpart. With respect to the spectral radius we observe that they are of order $O(N^{0.79})$, $O(N^{1.65})$ and $O(N^{2.56})$ for the first, second and respectively third order differentiation matrices.

4.4 The LC algorithm

Boundary conditions So far, we have been mainly concerned with homogeneous Dirichlet boundary conditions which have been introduced by deleting rows and columns (see Remark 1.6). With respect to boundary value problems formulated on infinite intervals these conditions can be *behavioral*, i.e., at infinity or *numerical*, i.e., in the origin (see the influential monograph [6] for this classification). The boundary conditions at infinity are directly satisfied by Laguerre functions. In fact we have

$$d^k p_{N-1}(\eta)/d\eta^k \to 0, \ \eta \to \pm\infty, \tag{4.17}$$

for any natural number k.

Now, an arbitrary number of m homogeneous or nonhomogeneous boundary conditions at zero (or at any other fixed real point) will be introduced by a *removing technique of independent boundary conditions* initiated in [15]. With this technique, we consider the boundary conditions on the origin as some linear independent constraints on the *state* of our system of N *degrees of freedom* and remove them since they are slaved to the other $N - m$ degrees which we keep. Thus, we decompose the state of the system into degrees of freedom that we keep and degrees of freedom that we remove using a *constraint matrix*. We express the removed state variables as a function of the kept ones, thus preserving their action in the system. Moreover, we are not interested in the *evolution* of the removed degrees of freedom since we can, at any time, recover them from the kept ones using a *give back* matrix.

Remark 4.1 In most cases, we can apply the procedure described above by simple matrix manipulations, the only two rules being:

1. For Dirichlet and Neumann boundary conditions, i.e., $m = 1$, remove the mesh point where the boundary condition applies.
2. For clamped boundary conditions (Dirichlet and Neumann at the same location) or hinged, mixed boundary conditions, when $m = 2$, remove the mesh points *at* and *next to* where the boundary condition applies.

A fairly similar technique is used in [41] where the authors introduce some hinged boundary conditions in a fourth order problem.

This simple and efficient algorithm in imposing the boundary conditions in the collocation method is a serious advantage when the method is compared with the Galerkin one.

The equations of the LC method The method casts the problem (4.13) into the following nonlinear algebraic system

$$\widetilde{\widetilde{LD}}^{(3)}\widetilde{\widetilde{\mathbf{F}}} + \left[\widetilde{\widetilde{LD}}^{(2)}\left(\widetilde{\widetilde{\mathbf{F}}}.\right)^2 + E_1\left(I + \widetilde{\widetilde{LD}}^{(2)}\right)\widetilde{\widetilde{\mathbf{F}}} + E_2\widetilde{\widetilde{LD}}^{(2)}\widetilde{\widetilde{\mathbf{F}}}\right]$$

$$-\beta\left[\left(\widetilde{\widetilde{LD}}^{(1)}\widetilde{\widetilde{\mathbf{F}}}\right).^2 + E_3\widetilde{\widetilde{LD}}^{(1)}\widetilde{\widetilde{\mathbf{F}}}\right] = \widetilde{\widetilde{\mathbf{g}}}. \tag{4.18}$$

In (4.18) the vector $\widetilde{\widetilde{\mathbf{F}}}$ contains the nodal unknown values F_j, $j = 3, ..., N$, i.e., the kept degrees of freedom from $\mathbf{F}$. The double tilde placed over differentiation matrices $\widetilde{\widetilde{LD}}^{(i)}$, $i = 1, 2, 3$ means that the first two rows and columns of the differentiation matrices $LD^{(n)}$, $n = 1, 2, 3$ are deleted and the Dirichlet and Neumann boundary conditions on zero were already incorporated using the constraint matrix. The matrices E_i, $i = 1, 2, 3$ are $(N - 2) \times (N - 2)$ diagonal matrices with the diagonal entries $e^{-\eta_k}$, $\eta_k - 1$ and $2\left(1 - e^{-\eta_k}\right)$, respectively, and $\widetilde{\widetilde{\mathbf{g}}}$ is the vector with the components

$$g_k = e^{-\eta_k}\left[2 - \eta_k - e^{-\eta_k} - \beta\left(2 - e^{-\eta_k}\right)\right], \quad k = 3, ..., N. \tag{4.19}$$

The MATLAB operators are used throughout this work and thus the elementwise power of a vector is denoted by $(\mathbf{F}.)^2$. The nonlinear algebraic system (4.18) was successfully solved by MATLAB built in function `fsolve`.

Recovering derivatives Thus, we first get $\widetilde{\widetilde{\mathbf{F}}}$ as a solution of (4.18), and then, using the give-back matrix we compute $\mathbf{F}$ and successively $\mathbf{F}' = LD^{(1)}\mathbf{F}$ and $\mathbf{F}'' = LD^{(1)}\mathbf{F}'$. Equation (4.12) produces the solution of the original problem. On the precision of this recovering process we will elaborate more in the next Section.

4.5 Numerical Solutions to Falkner-Skan Problem

A non-trivial test of accuracy First, we want to test the accuracy of our algorithm for nonlinear problem (4.8, 4.9). This can be accomplished upon introducing a modified source term $\widetilde{g}(\eta)$ in the r.h.s. of (4.13) such that this problem admits the exact solution $\eta^2 e^{-\eta}$. The result of integration for f, f' and f'' was exact to a precision of order 10^{-3}. As this error persists for vanishing and non vanishing β we can assess that the nonlinearity ff'' is overwhelming responsible, comparing with $(f')^2$, for this modest accuracy. An explanation can consist in the worse conditioning of the second order differentiation matrix with respect to the first one. We have also to mention that solving the nonlinear algebraic systems throughout this Section, the procedure was convergent in each and every situation.

Falkner-Skan and some particular problems The problem (4.8, 4.9) with $\beta = 0$ is the so called *Blasius problem*. Its solutions f, f', and f'' are depicted in Fig. 4.4. The first derivative f' has the physical meaning of mean tangential velocity as a function of normal distance.

The value of the second derivative $f''(0)$, which physically means the drag on the flat plate was found to be 0.4907. This value is larger than 0.33206 frequently

Fig. 4.4 Solution to Blasius problem when $N = 80$ and scaling factor equals 5

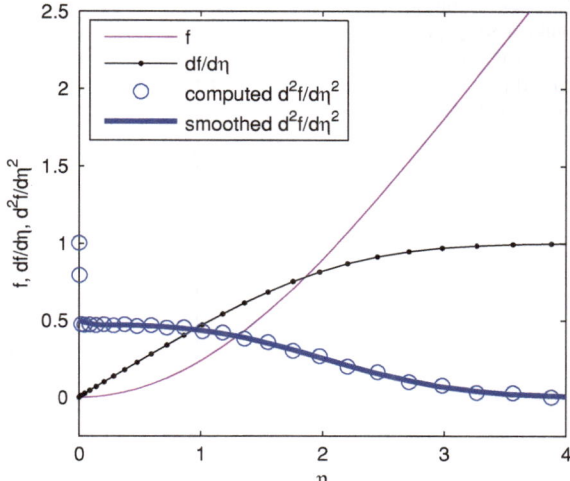

reported in literature and obtained by making use of shooting and domain truncation or 0.469601 reported in [30] and obtained by Fourier series. The special case $\beta = 1/2$ is called *Homann's equation*. In this case we have found $f''(0) = 0.8714...$ which is inferior to the value $0.9276...$ provided in [9]. The special case $\beta = 1$ is called *Hiemenz's equation* and corresponds to stagnation flow, for instance, past a large disk. In this case we have found $f''(0) = 1.1024...$ again inferior to $1.2325...$ furnished in the above quoted work.

With respect to our numerical results, we have to observe that very close to the flat plate the recovered values of the second derivative display a sort of numerical instability. A reason for this inconvenience can be the lack of a boundary condition for this derivative and again the round off errors which affects the second order differentiation matrix. However, with a usual cubic smoothing spline process over all recovered values of the second derivative, the first two noisy values can be removed and replaced by physically reasonable ones. This smoothing process is numerically stable with respect to N, i.e., $64 \leq N \leq 96$ and scaling factor b in the range [2,12] and is carried out with MATLAB built in function `spline`.

However, the mean velocity profile in Fig. 4.5, for positive (accelerating boundary layer flow) and negative β (decelerating boundary layer flow), shows excellent agreement with existing results in literature (see for instance [26, 32, 33]). If $\beta < 0$, so that the main stream is decreasing with respect to the longitudinal coordinate, there are two solutions of Falkner-Skan problem, provided that β is not less than -0.19884. One solution has a velocity profile of the normal 'kind', i.e., $0 \leq f'(\eta) \leq 1$, and the other has a region of reversed flow near the flat plate. A detailed discussion of the non-existence for β under this threshold is available in the monograph [1]. In our numerical experiments, for the same initial guess, i.e., equals **ones**$(N - 2, 1)$, we have obtained all normal 'kind' solutions in Fig. 4.5.

Fig. 4.5 Solutions to
Falkner-Skan problem for
$\beta = -0.0408, 0, 0.2$. The
cut off parameter N equals 80
and scale factor equals 5

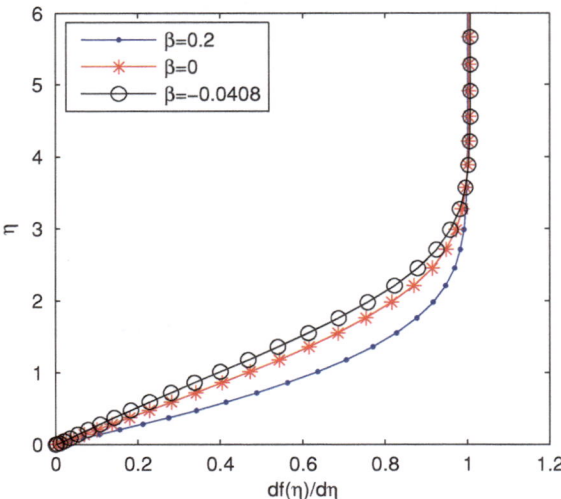

Remark 4.2 In the truncated boundary approach strategy, the boundary condition at infinity is replaced by the same condition at a given finite value L. Thus one can introduce $f_L(\eta)$ as the solution of the following problem

$$\begin{cases} f_L''' + f_L f_L'' = 0, \ 0 < \eta < L, \\ f_L(0) = f_L'(0) = 0, \ f_L'(L) = 1. \end{cases} \tag{4.20}$$

The error $e(\eta)$ related to $f_L(\eta)$ is defined by

$$e(\eta) := |f(\eta) - f_L(\eta)|, \ 0 \le \eta \le L. \tag{4.21}$$

In [31] the following upper bound for this error is provided

$$e(\eta) \le L f_L''(L) \, [f_L(L)]^{-1}. \tag{4.22}$$

It means that LC method used to solve this problem in our paper [11] has produced errors of order $O\left(10^{-3}\right)$ (see also Fig. 4.5). Roughly $L \, [f_L(L)]^{-1}$ approaches unity as $\eta \to L$ and thus the r. h. s. in (4.22) depends essentially on the drag at large distances away from fixed boundary.

We have to remark that in the actual computed value of $f_L(\eta)$, and its first two derivatives in L, the truncation and round off errors of the collocation are also incorporated. However, as we were not aware of the estimation (4.22) at the time the paper [11] was edited, this discussion validate once more the LC results.

4.6 Second Order Nonlinear Singular Boundary Value Problems on the Half-Line

Unsteady flow of a gas through a semi-infinite porous medium In [4, 5] the existence and uniqueness of positive solutions to some singular boundary value problems on finite and infinite intervals is established. We will consider such an example for a semi-infinite porous medium initially filled with gas at a uniform pressure. After a similarity transformation, the equation for the pressure of the gas in this medium reads

$$u''(z) = -\frac{2zu'(z)}{(1 - \alpha u(z))^{1/2}}, \ u(0) = 1, \ u \to 0, \ z \to \infty; \ 0 < \alpha \le 1.$$

For $\alpha < 1$ the problem is not singular except insofar as a semi-infinite interval is involved. However, on physical grounds a solution should exists when $\alpha = 1$ (diffusion into a vacuum) even though the differential equation is singular at origin. Using LC method we have successfully solve this problem for various values of the physical parameter α. The results are displayed in Fig. 4.6. It is visible from this picture that the solutions are monotone decreasing on $[0, \infty)$ and the slope increases as $\alpha \to 1$. Thus we confirm numerically the behavior predicted theoretically in both papers quoted above.

In our recent paper [12] we have considered the following two problems.

Cohen, Fokas, Lagerstrom model The first genuinely nonlinear problem was introduced by one of the authors of [7], namely Lagerstrom (1961)

$$\begin{cases} \frac{d^2u}{dx^2} + \frac{k}{x}\frac{du}{dx} + \alpha u\frac{du}{dx} + \beta \left(\frac{du}{dx}\right)^2 = 0, \ x \in (\varepsilon, \infty), \\ u = 0 \ at \ x = \varepsilon > 0, \ u \to U, \ x \to \infty. \end{cases} \tag{4.23}$$

The differential equation in (4.23) is the equation for the time independent temperature distribution in an infinite medium. Generally, it is somewhat an unrealistic physical model where x is a radial coordinate in $(k + 1)$-dimensional space and u is the temperature. The first two terms come from Laplace operator and the last two represent nonlinear autonomous heat sources unfortunately not validated in any actual physical model. The temperature equals zero on the sphere $x = \varepsilon, 0 < \varepsilon \ll 1$, and equals U at large distances. Using a *formal series expansion technique* Lagerstrom (1961) constructed asymptotic solutions for two $(k = 1)$ and three $(k = 2)$ dimensions, for $\beta = 0$ and $\beta = 1$. The intuitive reasoning indicated that for one-dimensional case $(k = 0)$ the problem is not singular. This was verified by the construction of an exact solution in this case (by quadrature and inversion of a function). If $\alpha > 0$ and $\beta > 0$ one can make their values unity (even simultaneously) by a scale transformation in u and x. However, because $\beta = 0$ is an interesting case and $\alpha = 0$ occurs in an auxiliary equation, i.e., Stokes equation, we consider arbitrary nonnegative values for α and β. For $\alpha > 0$, the problem (4.23) has a unique solution.

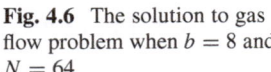

Fig. 4.6 The solution to gas
flow problem when $b = 8$ and
$N = 64$

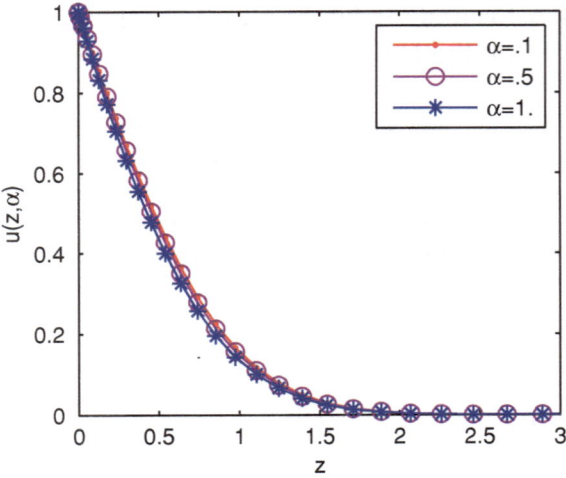

Fig. 4.7 The solution to the
Cohen, Fokas and Lagerstrom
problem when $b = 6$ and
$N = 64$

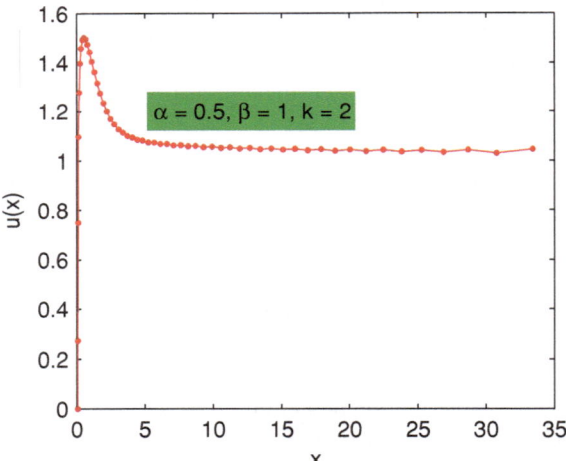

Our LC approximation for such a solution when $\varepsilon = 0.1$ and $U = 1$ is depicted in
Fig. 4.7.

Singular asymptotic techniques may and have been used only if $k \geq 1$. For $\alpha = 0$
the problem may be solved explicitly. The boundary value problem then has a solution
which is unique only if $k > 1$. Thus $k = 1$ is an important limiting case.

In [16] a rigorous discussion of the existence of a solution to (4.23) for $\beta = 0$, $k =
1$, and $\varepsilon \to 0^+$ is provided. Such a solution is depicted in Fig. 4.8 for the same value
of ε, U, N, and b as stated above. The case $\beta = 0$, $k = 1$ has the advantage
of being applicable to "real" problems, i.e., to some problems in fluid dynamics for
which (4.23) could be a model.

Fig. 4.8 A "real" solution
to the Cohen, Fokas and
Lagerstrom problem

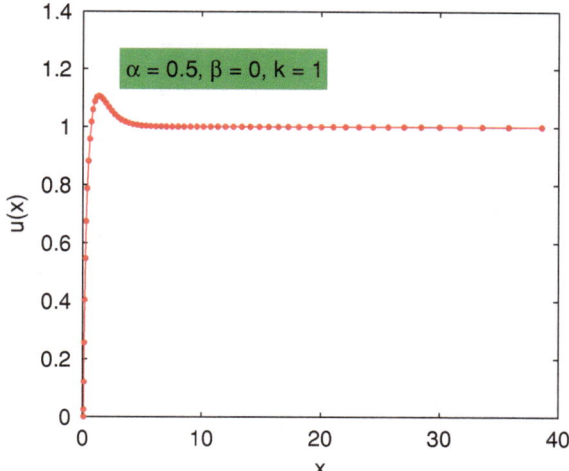

Both curves depicted in Figs. 4.7 and 4.8 behave in perfect accordance with the
asymptotic for $x \to \infty$ predicted in [24]. There is still a difference between them
for large x. It seems that the nonlinearity $\beta \left(\frac{du}{dx}\right)^2$ affects to some extent the stability
of the LC numerical process but ill conditioning of the differentiation matrices does
not appear to be a serious problem as our experiments indicate.

Solutions to density profile equation We investigate the *monotonously increasing solutions* of the problem

$$\begin{cases} \rho''(r) + \frac{N-1}{r}\rho'(r) = 4\lambda\,(\rho(r)+1)\,\rho(r)\,(\rho(r)-\xi),\, 0 < r < \infty, \\ \rho'(0) = 0,\ \rho \to \xi,\ r \to \infty, \end{cases} \tag{4.24}$$

where $\rho(r)$ stands for the density of a fluid. For the sake of simplicity we can choose
the following values of the parameters: $\lambda = 1$ without restriction of generality, $N = 3$
which corresponds to a physically meaningful case and ξ varying in the range $(0,1)$
such as to reflect different physical situations. The above equation is called the *density profile equation* and has the origins in the Cahn-Hillard theory which is used
in hydrodynamics to study the behavior of non-homogeneous fluids. In [18] the
authors find by polynomial collocation the so called "bubble-type solution". Analytical aspects concerning this equation, i.e., the existence and uniqueness of strictly
increasing solutions, their asymptotic behavior at infinity etc., as well as various
numerical solutions were thoroughly carried out in the series of papers pagination
[19, 20, 22]. When such a solution exists it has exactly one zero R in $(0, \infty)$,
and R is interpreted as the bubble radius. They also satisfy $-1 < \rho(r) < \xi$ and
$-1 < \rho(0) < 0$. The derivative of the solution attains a maximum at some value
$\tilde{r} < R$, and tends to 0 at infinity. Eventually, a bubble-type solution exhibits an *interior layer* which becomes sharper as $\xi \to 1$. We also have to observe that the problem

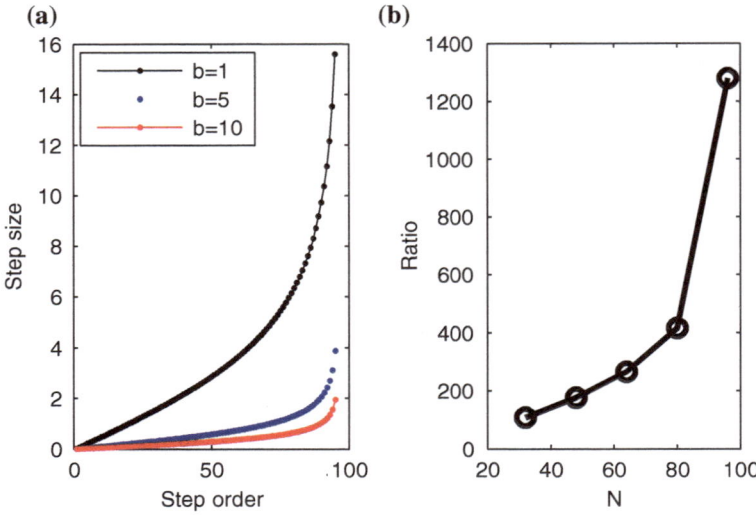

Fig. 4.9 The repartition of Laguerre nodes and step sizes for various N and b

(4.24) always admits the constant solution $\rho(r) = \xi$ which physically corresponds to the case of a homogeneous fluid, i.e., without bubbles.

In [18] the authors transform the problem to a finite interval and show that essential singular problem obtained is well-posed. In contrast, we solve (4.24) by LC. In fact, with respect to singularly perturbed problems, in [38], the authors observe that "for good resolution of the numerical solution at least one of the collocation points ought to lie in the boundary layer". Thus, the repartition of the Laguerre collocation nodes is of the utmost importance. We have used the repartition plotted in Fig. 4.9. For the singular problem at hand it seems fairly well suited. More precisely, for the nodes introduced with LC scheme we define by i the order of the step $[x_{i-1}, x_i]$, $i = 3, 4, ..., N$ and by $x_i - x_{i-1}$ its step size (the length). Consequently in Fig. 4.9a we have the dependence of the step size on the step order for various values of the scaling factor. In Fig. 4.9b we represent the dependence of the ratio of the largest step size to the smallest one upon the cut off parameter of interest. We notice a high non-uniformity of used meshes which supports fine resolutions in narrow regions with rapid variations of solutions.

As a maximal value, the above defined ratio for $N = 96$ and scaling factor $b = 10$ attains $1.2790e + 03$.

The LC algorithm, after an elementary homogenization of boundary condition at infinity, casts the problem (4.24) into a nonlinear algebraic system of type (4.18).

Corresponding to $\xi = 0.17, 0.34, 0.51, 0.68$ and 0.85 the solutions to this system are displayed in Fig. 4.10. The values of unknown function ρ in 0, the radius R, the elapsed CPU time, the number of iterations and the number of function evaluations when the nonlinear algebraic systems are solved by MATLAB code `fsolve` are

Fig. 4.10 Solutions to problem (4.24) for $\xi = 0.17, 0.34, 0.51, 0.68$ and 0.85 when $N = 64$ and $b = 10$ (from *left* to *right*)

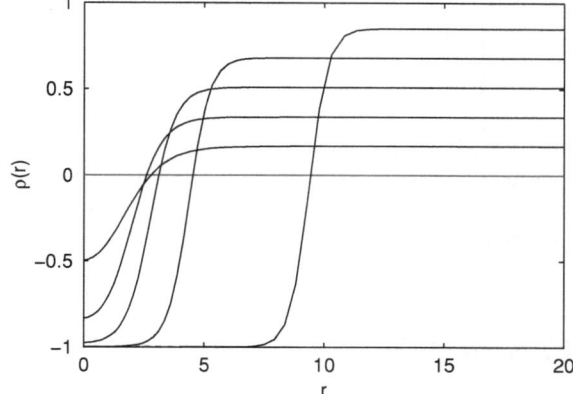

Table 4.1 Values of solutions in 0, the bubble radius, CPU elapsed time, number of iterations and number of function evaluations when $N = 64$ and $b = 10$

ξ	$\rho(0)$	R	CPU time/s	Iter	FuncEval.
0.17	−0.49436	2.7809	0.57812	8	576
0.34	−0.83219	2.5655	0.26562	9	514
0.51	−0.97509	3.0061	0.50000	15	898
0.68	−0.99952	4.5754	3.22815	998	63.621
0.85	−0.99999	9.5451	169.016	5240	334.983

reported in Table 4.1. All computations were carried out using 2010a variant of MATLAB on an HPxw8400 workstation with clock speed of 3.2 Ghz.

We consider that the solutions displayed in Fig.4.10 reproduce correctly the similar solutions reported in [18, 22] and satisfy all analytical properties theoretically established. The computed values of $\rho(0)$ and R from Table 4.1 are fairly closed to the corresponding values displayed in the above quoted papers. It is also worth nothing that the computational effort increases as parameter ξ approaches 1. It is apparent from the last three columns of Table 4.1. The same difficulty is reported in [18]. We have paid a particular attention to the assignment of the initial guesses. Thus for the first ξ the initial guess was a straight line joining the points $(0, -1)$ and $(x_N, 1)$. Then we have used a sort of continuation, i.e., the solution just found for a ξ becomes an initial guess for the next one.

As a final word we have to remark that our numerical experiments carried out with respect to the problem (4.24) and based on mapping coupled with ChC provided totally unsatisfactory results.

Remark 4.3 Up to our knowledge a general theory for the existence (and uniqueness) of solutions to problems (4.43) is not available. Thus, in spite of the fact that we do not want to elaborate in detail on this topic, we observe that for some nonlinearity f and boundary conditions the problem could be *embedded* in the following one

$$\begin{cases} \frac{1}{p(t)} \left(p(t) u'(t) \right)' = q(t) f\left(t, u(t), u'(t)\right), \ 0 < t < \infty, \\ u(0) = 0, \ u(t) \ bounded \ on \ [0, \infty), \end{cases} \quad (4.25)$$

where $f : [0, \infty) \times \mathbb{R} \times \mathbb{R} \to \mathbb{R}$ and $p, \frac{1}{q} : [0, \infty) \to [0, \infty)$ are assumed to be continuous. In [27] some existence results, establishing first the solution existence on a finite interval have been proved. Then they were extended on the semi-infinite interval by Arzela-Ascoli Theorem.

4.7 Second Order Eigenvalue Problems on Half-Line

In order to get more insight into the challenging problem (4.24) we will analyze its linearization around the constant solution $\rho(r) = \xi$. It reads

$$\begin{cases} \rho''(r) + \frac{N-1}{r} \rho'(r) = 4\lambda \left(\xi + \xi^2 \right) \rho(r), \ 0 < r < \infty, \\ \rho'(0) = 0, \ \rho \to \xi, \ r \to \infty. \end{cases} \quad (4.26)$$

The LC casts this problem into the following GEP

$$AX = \lambda BX, \quad (4.27)$$

where the matrices A and B are defined as

$$A := \widetilde{LD}^{(2)} + \mathbf{diag} \left(\frac{N-1}{\varsigma} \right) \widetilde{LD}^{(1)}, \ \ b := 4 \, \mathbf{diag} \left(\xi + \xi^2 \right),$$

the vector ς contains the nodes $x_2, x_3, ..., x_N$ and the matrices $\widetilde{LD}^{(i)}$, $i = 1, 2$ are defined in the manner exposed in Sect. 4.4. Therefore they incorporate the Neumann boundary condition in origin.

The pseudospectrum of this problem when $\xi = 0.4$ is provided in Fig. 4.11. Its outmost part slightly extends into the positive semi plane. On the other hand, the Henrici number, defined by (2.7), and computed for the matrix $A * B^{-1}$ is of order 1.070790. Both results suggest that the matrix pencil (A, B) is far from a normal one and consequently the linear stability analysis must be taken with some circumspection (see [39, 40]).

The eigenvalues are real and negative. The rightmost one was found to be $\lambda = -7.549523e - 004$ which implies the linear stability of constant solution.

The first five eigenvectors are available in Fig. 4.12.

All of them satisfy the boundary condition in the origin, are monotonously increasing for $0 < r < 20$ and approach with some decreasing oscillations the solution $\rho(r) = \xi$ for large r.

Fig. 4.11 The pseudospectrum of problem (4.26) when $b = 5$ and $N = 96$. The eigenvalues are marked with *dots*

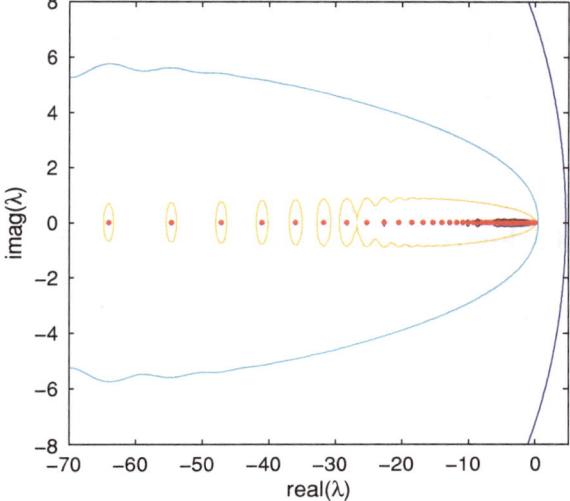

Fig. 4.12 The fist five eigenvectors (in ascending direction) of (4.26).

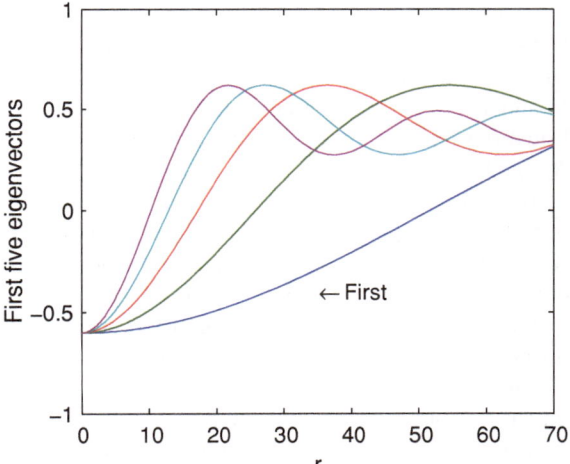

Throughout this work we have closely observed the eight *Rule-of-Thumb* formulated in the monograph [6]. Thus in order to test the reliability of our numerical results we repeated our numerical experiments with different N and have compared the results. Actually, in our LC scheme we have considered N in the range [64, 96].

A final remark is now opportune. Up to our knowledge there are some mathematical software packages devoted to singular S-L problems. SLEDGE is one of the most important and is described and tested in a series of papers (see for instance [28]). Beyond the computation of the eigenvalues and eigenfunctions this code attempts to classify each problem as to whether it is regular or singular, limit point or limit circle, oscillatory or non oscillatory etc. The code is essentially based on shooting and consequently on an empirical approximation of boundary condition at infinity. On the same techniques is based the code in [29]. The LC strategy introduced above is conceptually different and tested on some non-trivial examples from these papers provided reasonable solutions.

4.8 Fourth Order Eigenvalue Problems on Half-Line

Ekman boundary layer equations In [2] the authors consider the following GEP

$$\mathbf{M}^2\phi - i\gamma R_o \left(\widetilde{U} - c\right)\mathbf{M}\phi + i\gamma R_o \widetilde{U}_{zz}\phi + 2\psi_z = 0,$$
$$\mathbf{M}\psi - i\gamma R_o \left(\widetilde{U} - c\right)\psi - i\gamma R_o \widetilde{V}_z\phi - 2\phi_z = 0, \tag{4.28}$$

where $\mathbf{M}\phi := \phi_{zz} - \gamma^2 E_k \phi$, supplied with the boundary conditions

$$\phi(0) = \psi(0) = \phi_z(0) = 0, \ \phi, \ \phi_{zz}, \ \psi \to 0, \ z \to \infty. \tag{4.29}$$

If E_k is set equal to unity, the Rossby number R_o equals the Reynolds number R_e and these equations are reduced exactly to those of [23]. Here γ is the wave number, $\widetilde{U}$ and $\widetilde{V}$ are two known and smooth functions and c is the complex phase speed. Equations (4.28, 4.29) define an eigenvalue problem which determines c in terms of R_o, γ and E_k.

In [14] the authors carried out a detailed analysis of this particular case, from operator theory point of view. They locate the essential spectrum and analyze the L^2 solutions of the factorized system of first order differential equations. Eventually, by domain truncation and shooting they solved numerically this system.

One important remark is now appropriate. The system (4.28, 4.29) can be transformed into a first order differential system of six equations as in [2, 14]. The converse statement is also true, i.e. eliminating for instance ϕ variable, the system can be cast into a sixth order differential equation. As the algebraic manipulations in the general case are tedious, we consider only the case when $\gamma = 0$. In this case we get the polynomial eigenvalue problem

$$-\psi^{(vi)} - 4\psi'' = 2\lambda\psi^{(iv)} + \lambda^2\psi'', \ z \in (0, \infty),$$
$$\psi(0) = \psi'(0) = 0, \ \psi^{(v)}(0) + 4\psi'(0) + \lambda\psi'''(0) = 0, \ \psi, \ \psi'' \to 0, \ z \to \infty, \tag{4.30}$$

with boundary conditions depending on the spectral parameter. Unfortunately, from computational point of view this formulation does not open feasible perspectives.

LC solutions Rather than reproduce all the results from [2, 23] etc. we just consider one example from these papers, namely the flow profiles

$$\widetilde{U}(z) := \cos\varepsilon - \exp(-z)\cos(z + \varepsilon), \ z \in [0, \infty),$$
$$\widetilde{V}(z) := -\sin\varepsilon + \exp(-z)\sin(z + \varepsilon), \ z \in [0, \infty), \ |\varepsilon| < 1. \tag{4.31}$$

The dependence of the leftmost eigenvalue λ on the small parameter ε, the Reynolds number R_e, and γ, is actually studied for the following LC discretization of the problem (4.28, 4.29),

$$AX = \lambda BX, \tag{4.32}$$

where

$$A := \left(\begin{matrix} \left(-\widetilde{LD}^{(2)} + \gamma^2\right)^2 + i\gamma\,R_e\,\tilde{V}\left(-\widetilde{LD}^{(2)} + \gamma^2\right) + i\gamma\,R_e\,\tilde{V}'' & 2\widetilde{LD}^{(1)} \\ 2\widetilde{LD}^{(1)} + i\gamma\,R_e\,\tilde{U}' & -\widetilde{LD}^{(2)} + \gamma^2 + i\gamma\,R_e\,\tilde{V} \end{matrix} \right), \tag{4.33}$$

and

$$B := \left(\begin{matrix} -\widetilde{LD}^{(2)} + \gamma^2 & 0 \\ 0 & I \end{matrix} \right). \tag{4.34}$$

In this GEP, $X := \left(\phi\ \psi \right)^T$, ϕ and ψ being the unknown vectors of dimension $N-2$ and

$$\lambda := i\gamma\,R_e c. \tag{4.35}$$

The matrices $\widetilde{LD}^{(i)}$, $i = 1, 2$ are the differentiation matrices with the boundary conditions (4.29) introduced as in Sect. 4.4. Along with the pencil (A, B) above, we will study the pencil (A_0, B) which provides surprising information for the spectrum of GEP (4.32). The matrix A_0 reads

$$A_0 := \left(\begin{matrix} \left(-\widetilde{LD}^{(2)} + \gamma^2\right)^2 & 2\widetilde{LD}^{(1)} \\ 2\widetilde{LD}^{(1)} & -\widetilde{LD}^{(2)} \end{matrix} \right), \tag{4.36}$$

and is obtained from the matrix A making $R_e = 0$.

In [25] the GEP (4.32) is solved as an optimization problem and the author finds that for the values of parameters

$$\gamma = 0.316225,\ \varepsilon = -23.326108^0,\ R_e = 54.155038, \tag{4.37}$$

the corresponding eigenvalue c has a vanishing imaginary part and the real part equals 0.616301. For the same values of the above parameters, the leftmost eigenvalue we have obtained reads

$$c = 5.563982e - 01 - i2.821821e - 02, \tag{4.38}$$

when the problem (4.32) was solved with $N = 80$ and $b = 4$. In fact, we have obtained values of the leftmost c which are indistinguishable up to the sixth digit, for

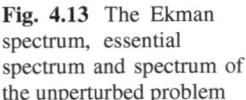

Fig. 4.13 The Ekman spectrum, essential spectrum and spectrum of the unperturbed problem

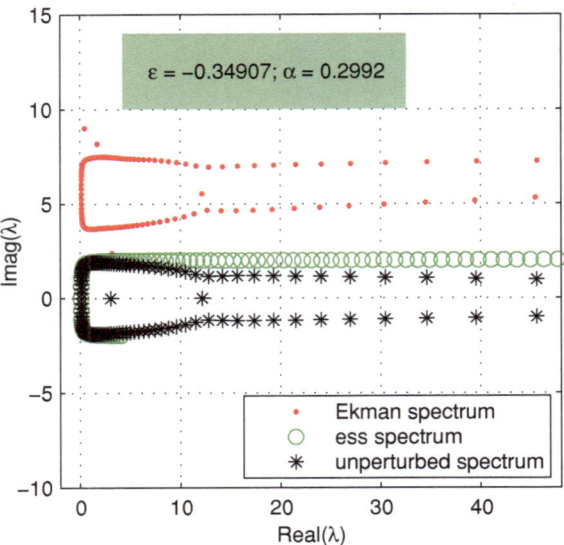

N in the range [56,96] and scaling factor in the range [2,12]. This means numerical stability of our numerical process. These numerical results are also in reasonable agreement with those published in [2, 17, 23].

In order to find the entire spectrum of the GEP we have used the `eig` MATLAB code and the accuracy of a limited number of eigenvalues from the leftmost region of this spectrum was validated using the JD algorithm (see our paper [13]).

With the aim to validate our LC results, and to compare them with the theoretical ones from [14], we have displayed the left part of the spectrum of the *unperturbed symmetric* pencil (A_0, B) and of the spectrum of (A, B) for values of parameters from (4.37), in Fig. 4.13.

The *essential spectrum* is indeed contained inside the predicted curve given parametrically by

$$\lambda(t) = t^2 + \gamma^2 + \frac{2it}{\sqrt{t^2 + \gamma^2}}, \qquad (4.39)$$

where t is a real parameter and i the imaginary unit. The curve is marked with circles on Fig. 4.13.

However, the spectrum of the GEP (4.32) appears to be a slightly perturbation of the symmetric spectrum of (A_0, B), i.e., the leftmost part is slightly distorted and three 'rebel' eigenvalues emerge. More than that, with Theorem 5.1 in [14], the authors show that the essential spectrum and the spectrum of the Ekman problem locate in a semi-infinite strip. Denoting $\lambda := \nu + i\mu$, this theorem predicts, in the case at hand, that $\nu \geq -1.355866e + 02$ and $|\mu| \leq 4.599409e + 02$. Our numerical results infer that these bounds overestimate the real situation.

Fig. 4.14 The first
eigenmodes of problem (4.32)

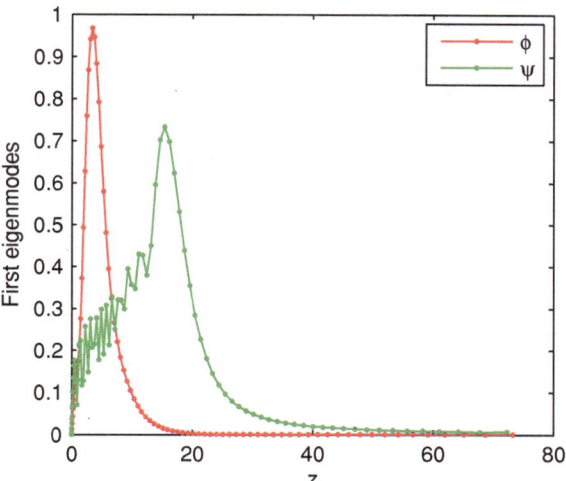

Moreover, the first two eigenmodes ϕ, ψ are depicted in Fig. 4.14. They behave correctly at the ends of integration interval.

We have also computed numerically the pseudospectrum of (4.32) for the values (4.37) of parameters γ, ε and R_e. It is reported in Fig. 4.15 and suggests that the pencil (A, B) in this *marginal case* is *non-normal* [see (2.9)].

Thus, according with the ideas exposed in [39, 40] this linear hydrodynamic stability analysis has to be interpreted with some precaution. In other words, the pseudospectrum of the involved algebraic GEP suggests the limitations of this analysis because it slightly extends into the left hand side semi plane.

Remark 4.4 With respect to the computational cost of the LC algorithm we have to notice the following. Once the Laguerre differentiation matrices are accurately provided by a built in function the rest of operations is carried out at a fairly low cost. The introduction of *any type of linear boundary condition* in zero is carried out by some elementary manipulations of the constraint matrix. Then a boundary value problem requires the solution of a nonlinear algebraic system and an eigenvalue one implies at least a specified region of the spectrum of a GEP. Some elementary operations with the give back matrix produce the removed degrees of freedom. The optimal value of scaling parameter has to be adjusted manually. This could be a time-consuming process which asks some computing experience and could be considered as a drawback of algorithm. However, the MATLAB built in code $\texttt{fsolve}$ as a solver for nonlinear algebraic systems, and respectively $\texttt{eig}$ or any of the JD solvers for GEPs, have worked smoothly without major incidents (non-convergent outcomes). The elapsed CPU times for all our numerical experiments have been fairly resonable.

Remark 4.5 The removing technique from Sect. 4.4 was successful in introducing the largest imaginable variety of linear boundary conditions including *slip boundary*

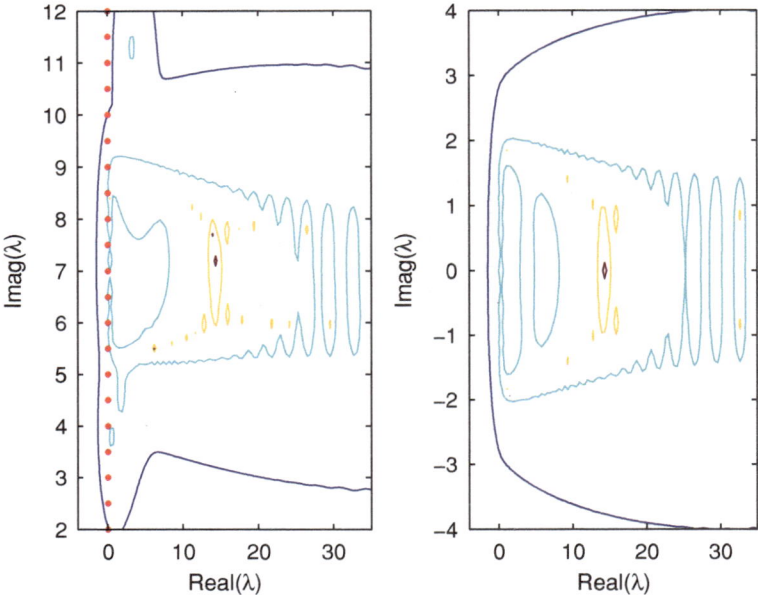

Fig. 4.15 The pseudospectrum of problem (4.32) (*left picture*) and of the unperturbed problem, when $N = 96$ and $b = 8$

conditions. Thus, we have confirmed the results of [37]. In this chapter the authors recently solved a Poiseuille flow stability problem in porous medium which involves such boundary conditions.

4.9 The Movement of a Pile

Let $u(x)$ be the deflection of a semi-infinite pile embedded in soft soil at a distance x below the surface of the soil. The governing differential equation for the movement of the pile, in dimensionless form, is given by

$$\frac{d^4 u}{dx^4} = -P_1 \left(1 - \exp(-P_2 u)\right), \ 0 < x < \infty, \tag{4.40}$$

where P_1 and P_2 are positive material constants. At the origin, a zero moment and a positive shear P_3 are assumed, i.e.,

$$\frac{d^2 u}{dx^2}(0) = 0, \ \frac{d^3 u}{dx^3}(0) = P_3. \tag{4.41}$$

Moreover, from physical considerations it follows that $u(x)$ and all its derivatives go to zero at infinity, so that the following behavioral (asymptotic) boundary conditions

$$u, \frac{du}{dx} \rightarrow 0, \ x \rightarrow \infty, \tag{4.42}$$

can be imposed. The boundary value problem (4.40)–(4.42) is of interest in foundation engineering; for instance, in the design of drilling rigs above the ocean floor.

In order to solve this problem numerically we introduce the new unknown $w(x)$ by

$$u(x) := w(x) + h(x),$$

where $h(x) := x^3 e^{-x} P_3/6$. Thus, we actually solve the *homogeneous nonlinear boundary value problem*

$$\begin{cases} \frac{d^4 w}{dx^4} = -P_1 \left(1 - \exp(-P_2 (w + h))\right) - \frac{d^4 h}{dx^4}, \ 0 < x < \infty, \\ \frac{d^2 w}{dx^2} (0) = 0 = \frac{d^3 w}{dx^3} (0), \ w, \ \frac{dw}{dx} \rightarrow 0, \ x \rightarrow \infty. \end{cases} \tag{4.43}$$

In [8] the author solves the problem (4.40)–(4.42) by a method based on compound matrix factorization and truncates the $[0, \infty)$ domain to a finite one. Then he compares his results with those reported in [21] where the problem is considered as a test one in working with the box difference scheme.

A theory for defining asymptotic boundary conditions to be imposed at the end of the truncated domain has been developed by many scientists but it is still unsatisfactory.

In order to avoid this serious difficulty we solve the above problem by LC method. We mention again that the important advantage of this method resides in the fact that the involved Laguerre functions automatically satisfy the boundary conditions at infinity (4.42). Details on the implementation of this method are available in our recent paper [12].

However, very sketchy, we solve numerically the non-homogeneous nonlinear algebraic system

$$\widetilde{\widetilde{LD}}^{(4)} \widetilde{\widetilde{w}} = -P_1 \left(I - \exp\left(-P_2 \left(\widetilde{\widetilde{w}} + \mathbf{h}\right)\right)\right) - \widetilde{\widetilde{LD}}^{(4)} \mathbf{h}, \tag{4.44}$$

where the matrix $\widetilde{\widetilde{LD}}^{(4)}$ stands for the fourth order differentiation matrix with the boundary conditions (4.41) enforced and the vectors $\widetilde{\widetilde{w}}$, $\mathbf{h}$ stand respectively for the unknown vector and the vector of values of function h on the nodes x_i, $i = 3, 4, ..., N$. We mention that due to the fact that we have two independent boundary conditions in (4.41) we have to remove the first two nodes $x_1 = 0$ and x_2. Thus the system (4.44) is of order $N - 2$.

We have solved this nonlinear system starting from the initial guess, the vector **ones** of order $N - 2$. The MATLAB routine `fsolve` has used only 17 iterations and 762 function evaluations in order to get a convergent solution. Variations of this

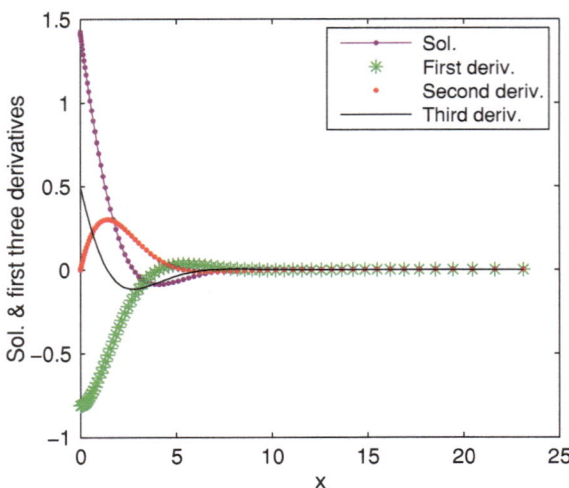

Fig. 4.16 The Laguerre collocation solution to problem (4.40)–(4.42) along with its first three derivatives when $N = 76$ and $b = 12$

initial guess, of the cut off parameter N and of the scaling factor b have led to the same solution.

Thus, we first get $\widetilde{\widetilde{\mathbf{w}}}$ as a solution of (4.44), and then, using the give-back matrix we compute the entire vector $\mathbf{w}$. Further on, by successive multiplications with first order differentiation matrix, $\mathbf{w}' = LD^{(1)}\mathbf{w}$ etc., we have recovered its derivatives of first, second and third orders (see Sect. 4.4).

The LC solution, along with its first three derivatives, to problem (4.40)–(4.42) for the values of parameters

$$P_1 = 1, \ P_2 = \frac{1}{2}, \ P_3 = \frac{1}{2},$$

are depicted in Fig. 4.16. They seem to be in perfect qualitative accordance with those presented in [8].

References

1. Acheson, D.J.: Elementary Fluid Dynamics. Clarendon Press, Oxford (1992)
2. Allen, L., Bridges, T.J.: Hydrodynamic stability of the Ekman boundary layer including interaction with a compliant surface: a numerical framework. Eur. J. Mech. B Fluids **22**, 239–258 (2003)
3. Bernardi, C., Maday, Y.: Approximations Spectrales de Problems aux Limites Elliptiques. Springer, Paris (1992)
4. Baxley, J.V.: Existence and uniqueness for nonlinear boundary value problems on infinite intervals. J. Math. Anal. Appl. **147**, 122–133 (1990)
5. Bobisud, L.E.: Existence of positive solutions to some nonlinear singular boundary value problems on finite and infinite intervals. J. Math. Anal. Appl. **173**, 69–83 (1993)

6. Boyd, J.P.: Chebyshev and Fourier Spectral Methods, 2nd edn. Dover Publications, Inc., Mineola (2000)
7. Cohen, D.S., Fokas, A., Lagerstrom, P.A.: Proof of some asymptotic results for a model equation for low Reynolds number flow. SIAM J. Appl. Math. **35**, 187–207 (1978)
8. Fazio, R.: A free boundary value approach and Keller's box scheme for BVPs on infinite intervals. Int. J. Comput. Math. **80**, 1549–1560 (2003)
9. Finch, S.: Prandtl-Blasius flow. http://algo.inria.fr/csolve/bla.pdf (2008/11/12). Accessed 12 June 2012
10. Fornberg, B.: A Practical Guide to Pseudospectral Mathods. Cambridge University Press, Cambridge (1998)
11. Gheorghiu, C.I.: Laguerre collocation solutions to boundary layer type problems. Numer. Algor. **64**, 385–401 (2013)
12. Gheorghiu, C.I.: Pseudospectral solutions to some singular nonlinear BVPs. Applications in Nonlinear Mechanics. Numer. Algor. doi:10.1007/s11075-014-9834-z.
13. Gheorghiu, C.I., Rommes, J.: Application of the jacobi-davidson method to accurate analysis of singular linear hydrodynamic stability problems. Int. J. Numer. Method Fluids **71**, 358–369 (2012)
14. Greenberg, L., Marletta, M.: Numerical methods for higher order Sturm-Liouville problems. J. Comput. Appl. Math. **125**, 367–383 (2000)
15. Hoepffner, J.: Implementation of boundary conditions. http://www.lmm.jussieu.fr/hoepffner/boundarycondition.pdf (2010). Accessed 25 Aug 2012
16. Hsiao, G.C.: Singular perturbations for a nonlinear differential equation with a small parameter. SIAM J. Math. Anal. **4**, 283–301 (1973)
17. Ioss, G., Bruun, H.: True, bifurcation of the stationary Ekman flow into a stable periodic flow. Arch. Rat. Mech. Anal. **68**, 227–256 (1978)
18. Kitzhofer, G., Koch, O., Lima, P., Weinmüller, E.: Efficient numerical solution of the density profile equation in hydrodynamics. J. Sci. Comput. **32**, 411–424 (2007)
19. Konyukhova, N.B., Lima, P.M., Morgado, M.L., Soloviev, M.B.: Bubbles and droplets in nonlinear physics models: analysis and numerical simulation of singular nonlinear boundary value problems. Comput. Math. Math. Phys. **48**, 2018–2058 (2008)
20. Kulikov, G.Y., Lima, P.M., Morgado, M.L.: Analysis and numerical approximation of singular boundary value problems with p-Laplacians in fluid mechanics. J. Comput. Appl. Math. http://dx.doi.org/10.1016/j.cam.2013.09.071
21. Lentini, M., Keller, H.B.: Boundary value problems on semi-infinite intervals and their numerical solution. SIAM J. Numer. Anal. **17**, 577–604 (1980)
22. Lima, P.M., Konyukhova, N.B., Chemetov, N.V., Sukov, A.I.: Analytical- numerical investigation of bubble-type solutions of nonlinear singular problems. J. Comput. Appl. Math. **189**, 260–273 (2006)
23. Lilly, D.K.: On the instability of Ekman boundary flow. J. Atmospheric Sci. **23**, 481–494 (1966)
24. Markowich, P.A.: Analysis of boundary value problems on infinite intervals. SIAM J. Math. Anal. **14**, 11–37 (1983)
25. Melander, M.V.: An algorithmic approach to the linear stability of the Ekman layer. J. Fluid Mech. **132**, 283–293 (1983)
26. Ockendon, H., Ockendon, J.R.: Viscous Flow. Cambridge University Press, Cambridge (1995)
27. O'Regan, D.: Solvability of some singular boundary value problems on the semi-infinite interval. Can. J. Math. **48**, 143–158 (1996)
28. Pruess, S., Fulton, C.T.: Mathematical software for Sturm-Liouville problems. ACM Trans. Math. Softw. **19**, 360–376 (1993)
29. Pryce, J.D.: A test package for Sturm-Liouville solvers. ACM Trans. Math. Softw. **25**, 21–57 (1999)
30. Rosales-Vera, M., Valencia, A.: Solutions of Falkner-Skan equation with heat transfer by Fourier series. Int. Comm. Heat Mass Transf. **37**, 761–765 (2010)
31. Rubel, L.A.: An estimation of the error due to the truncated boundary in the numerical solution of the Blasius equation. Quart. Appl. Math. **13**, 203–206 (1955)

32. Schlichting, H.: Boundary Layer Theory, 4th edn. McGraw-Hill, New York (1960)
33. Schmid, P.J., Henningson, D.S.: Stability and Transition in Shear Flows. Springer, New York (2001)
34. Shen, J.: Stable and efficient spectral methods in unbounded domains using Laguerre functions. SIAM J. Numer. Anal. **38**, 1113–1133 (2000)
35. Shen, J., Wang, L.-L.: Some recent advances on spectral methods for unbounded domains. Commun. Comput. Phys. **5**, 195–241 (2009)
36. Shen, J., Tang, T., Wang, L.-L.: Spectral Methods. Springer, Berlin (2011)
37. Straughan, B., Harfash, A.J.: Instability in Poiseuille flow in a porous medium with slip boundary conditions. Microfluid. Nanofluid (2012). doi:10.1007/s10404-012-1131-3
38. Tang, T., Trummer, M.R.: Boundary layer resolving pseudospectral methods for singular perturbation problem. SIAM J. Sci. Comput. **17**, 430–438 (1996)
39. Trefethen, L.N.: Pseudospectra of linear operators. SIAM Rev. **39**, 383–406 (1997)
40. Trefethen, L.N., Trefethen, A.E., Reddy, S.C., Driscoll, T.A.: Hydrodynamic stability without eigenvalues. Science **261**, 578–584 (1993)
41. Weideman, J.A.C., Reddy, S.C.: A MATLAB differentiation matrix suite. ACM Trans. Math. Softw. **26**, 465–519 (2000)

Chapter 5
Conclusions and Further Developments

Abstract Spectral tau, Galerkin and collocation methods are briefly revised. On bounded domains all of them are constructed using Chebyshev polynomials. Collocation on an unbounded domain is based on Laguerre functions. Practical and computational aspects of these methods are mainly emphasized. High order eigenvalue problems, i.e., of fourth, sixth and eighth orders along with genuinely nonlinear and singular perturbed two-point eigenvalue problems are considered. The capabilities of the methods are analysed based on the conditioning and normality of the differentiation matrices in both the physical and phase (coefficient) spaces.

Keywords Boundary value problems · Collocation · Conditioning · Differentiation matrices · Eigenvalue problems · Galerkin · High order · Multiparameter · Nonlinear · Normality · Singularly perturbed · Tau

> *Applied mathematics is subject to fads and enthusiasms. We have shown above that the theory of numerical algorithms contains hidden beyond-all-orders terms, but this aspect of numerical analysis is largely terra incognita.*
> *John P. Boyd, The Devil's Invention: Asymptotic, Superasymptotic and Hyperasymptotic Series, 2000*

5.1 Lessons Learned Along the Way

It is largely accepted that in assessing the effectiveness of various algorithms we are concerned with the following attributes listed in decreasing order of importance: generality, reliability, stability, accuracy, efficiency, storage requirements, ease of use and simplicity. We advocate that spectral methods, as a whole, have all the attributes listed above except generality. Spectral methods based on Chebyshev polynomials have shown their capabilities in solving various fairly challenging differential problems on finite intervals. But they are not generally applicable. Every method, tau,

C.-I. Gheorghiu, *Spectral Methods for Non-Standard Eigenvalue Problems*, SpringerBriefs in Mathematics, DOI: 10.1007/978-3-319-06230-3_5, © The Author(s) 2014

collocation and Galerkin, solves efficiently and accurately only particular classes of problems.

Thus, with respect to the *tau* method we can conclude that it restricts its applicability to linear problems because nonlinearities are harder to deal in phase (coefficient) space due to the convolution they entail. However, the tau method produces derivatives in phase space and this is an important advantage. Moreover, it is the unique choice when a GEP contains boundary conditions involving the eigenparameter. We have exploited this fact in solving a non-standard O-S problem attached to the Marangoni-Plateau-Gibbs effect. Our numerical results were confirmed in independent studies. Unfortunately, to visualize a tau solution one needs a discrete transform (Fourier, Chebyshev etc.) in order to recover its nodal values.

The *collocation* method efficiently manipulates the nonlinearities in the physical (real) space and the unique concern with this computation refers to the accuracy in computing differentiation matrices. It was the best choice in resolving interior and boundary layers, i.e., in solving nonlinear singularly perturbed problems. It also provided excellent results in solving genuinely high order (of sixth and eighth orders) eigenvalue problems in conjunction with our "$D^{(2)}$" strategy. This strategy drastically reduced the differential order of the problem and of the boundary conditions as well. Moreover, the generated algebraic GEPs benefited from a better conditioning.

The ChC method has proved its reliability and accuracy in solving the Mathieu's system as a MEP. It is known that accurate numerical computation of high frequencies (eigenvalues) is much harder than for low frequencies. In solving this classical problem, we have introduced a hierarchy of numerical algorithms that can deal with the corresponding algebraic eigenvalue problems for increased orders of approximation N. Thus, we have been able to compute the eigenfrequencies and the corresponding eigenmodes (eigenvectors) from the first ones to the order of about 10^4. Accordingly, the accuracy varies from a quasi spectral one to a moderate one for the highest mode.

With respect to the accuracy as well as to the CPU time and storage required, our algorithm is superior to that based on FD method. It is also stable with respect to the order of approximation as it is apparent from our reported numerical experiments. Moreover, we have solved the MEP attached to Mathieu's system for domains with geometrical aspects ranging from a circle to extremely flattened ellipse, i.e., a ratio of the major to minor axes of ellipses of order 10^3. This means that the method is stable with respect to the geometry (eccentricity) of the problem. It is also of some importance to remark that, up to our knowledge, pseudospectra of Mathieu's system were for the first time reported in one of our previous papers as well as in this work.

For large algebraic GEPs and MEPs as well as for singular GEPs we have particularly dealt with the *JD* type methods. These subspace type methods are target oriented and consequently they eliminate spurious (at infinity) eigenvalues. We have estimated their cost (CPU time and storage required) and based on the pseudospectrum of the matrix pencil attached to a singular GEP, we have concluded on their convergence and stability in the computation of the first two eigenvalues. The $rJDQZ$ and Arnoldi are mainly compared and founded to be fairly accurate.

Whenever, test and trial functions satisfying all the essential boundary conditions can be built, the best (spectral) accuracy is guaranteed by *Galerkin* type methods. The sparsity, normality and conditioning of the differentiation, and consequently corresponding discretization matrices, strongly depend on the choice of these test and trial functions. With respect to the normality and conditioning, our numerical experiments proved that Galerkin schemes can provide the best discretization matrices. Collocation and tau methods provide full populated matrices the latter method being the worst with respect to both parameters. It is also interesting to remark that one basis, i.e., the Heinrichs' basis, proved to be fairly useful in working with all the three methods. Of course, the differentiation matrices involved appeared with fairly different qualities. Thus, if one ever finds himself/herself on the verge of using spectral method in his/her research, is strongly encouraged to look into how to deal with test and trial functions in order to effectively work with the best conditioned discretization matrices.

However, mainly with respect to collocation method, the topic of enforcing nonlinear boundary conditions or boundary conditions depending on the spectral parameter, when GEPs are considered is still open.

With respect to boundary value problems formulated on half-line we have been concerned with the *Laguerre collocation*. We recall that it was incorporated without any difficulty, and avoiding domain truncation, any boundary conditions at infinity, i.e., it enhanced the asymptotic behavior of solution at large distances. We also have to mention that for problems on infinite intervals we restrict ourselves to this method. In this context we have elaborated a short analysis on the influence of the scaling factor on the conditioning and normality of Laguerre differentiation matrices. These matrices remain fully populated but are better conditioned than their Chebyshev counterpart. Anyway, the behavior at infinity of solutions to linear problems which decay without oscillations at large distances is reproduced with a quasi spectral accuracy by the method.

It is well known that in order to obtain a fine resolution for fairly sharp layer problems, at least some of the grid points should lie in the boundary layer no matter how narrow this is. In the Laguerre collocation method discussed above this desiderate is accomplished by making use of a scaling parameter. Its tuning is carried out manually in a reduced number of computing experiments. However, this could be considered as a drawback of our strategy when compared with the automatic mesh adaptation for sharp layers proposed in some works. We have provided reliable numerical results for classical Falkner-Skan (Blasius) boundary layer problems, for monotonously increasing solutions to density profile equations, for the linear stability of Ekman boundary layer and for the governing problem of the deflections of a semi-infinite pile embedded in soft soil, to quote only the most important problem analyzed.

Throughout this work the normality of the differentiation (discretization) matrices has been quantified using a scalar measure, i.e., the Henrici's number and the pseudospectrum as a companion clue. This important characteristic of these matrices, along with their conditioning, shed some light on the capabilities of various spectral methods. Working with bases of Chebyshev polynomials the Galerkin type methods are the most normal. Collocation and tau methods come in decreasing order.

But with respect to the collocation method, Chebyshev bases produce more normal differentiation matrices than Laguerre functions. Thus we have separated at least partially the effect of the discretization and that of the choice of bases on the normality. Consequently, people interested in working with these methods have additional information in choosing a particular one.

As a side remark, we believe that numerical solutions to Cohen, Fokas and Lagerstrom type equation are for the first time obtained and reported.

5.2 Further Developments

An automatic generation of a suitable scaling parameter, for a specific problem formulated on an infinite domain, remains a tremendous important open issue. It will be considered in the future works.

At the same time, in order to optimize and extend the applicability of the Laguerre collocation in solving time dependent nonlinear problems a *Laguerre transform* between physical and phase spaces would be fairly necessary. Some work is in progress concerning this topic.

Appendix A
Algebraic Two-Parameter Eigenvalue Problems

A.1 Tensor Products of Matrices

First, we shortly review some tensorial algebra.

Definition 1.1 Given two square matrices $A := (a_{ij})$ and $B := (b_{ij})$ of dimension s and p respectively, we define the *tensor product* of A and B, $A \otimes B$, to be the $ps \times ps$ dimensional matrix

$$A \otimes B := \begin{pmatrix} a_{11}B & a_{12}B & \cdots & a_{1n}B \\ a_{21}B & a_{22}B & \cdots & a_{2n}B \\ \vdots & \vdots & \ddots & \vdots \\ a_{n1}B & a_{n2}B & \cdots & a_{nn}B \end{pmatrix}.$$

We collect some important properties of tensor product in the following Lemma.

Lemma 1.1 *Let A and B be two square matrices of dimension s and p respectively. Then, the following properties hold*

1. $(A \otimes B)^T = A^T \otimes B^T$,
2. $(A \otimes B)^{-1} = A^{-1} \otimes B^{-1}$,
3. *if A has eigenvalues λ_i, $i = 1, 2, \ldots, s$ and B has eigenvalues μ_j, $j = 1, 2, \ldots, p$ then $A \otimes B$ has eigenvalues $\lambda_i \mu_j$, $i = 1, 2, \ldots, s$ and $j = 1, 2, \ldots, p$;*
4. *if $A, C \in \mathbb{R}^{s \times s}$ and $B, D \in \mathbb{R}^{p \times p}$ then $(A \otimes B)(C \otimes D) = AC \otimes BD$.*

A.2 Algebraic Two-Parameter Eigenvalue Problems

As we have considered in Sect. 3.5 the differential Mathieu's system as a MEP, we briefly review now the concept of algebraic two-parameter eigenvalue problem.

C.-I. Gheorghiu, *Spectral Methods for Non-Standard Eigenvalue Problems*, SpringerBriefs in Mathematics, DOI: 10.1007/978-3-319-06230-3, © The Author(s) 2014

An *algebraic two-parameter eigenvalue problem* has the form

$$\begin{cases} A_1\mathbf{x} = \lambda\, B_1\mathbf{x} + \mu\, C_1\mathbf{x}, \\ A_2\mathbf{y} = \lambda\, B_2\mathbf{y} + \mu\, C_2\mathbf{y}, \end{cases} \tag{A.1}$$

where A_i, B_i, and C_i are given $n_i \times n_i$ complex matrices, λ, μ are complex, $\mathbf{x}$, and $\mathbf{y}$ are complex vectors with n_1 and respectively n_2 components. A pair (λ, μ) is called an *eigenvalue* if it satisfies (A.1) for non-zero vectors $\mathbf{x}$ and $\mathbf{y}$. The tensor product $x \otimes y$ is then the corresponding *eigenvector*.

The two-parameter eigenvalue problem (A.1) can be expressed as two coupled GPEs as follows. On the tensor product space $S := \mathbb{C}^{n_1} \otimes \mathbb{C}^{n_2}$ of the dimension $m := n_1 n_2$ we define the so called *operator determinants*

$$\Delta_0 = B_1 \otimes C_2 - C_1 \otimes B_2, \tag{A.2}$$

$$\Delta_1 = A_1 \otimes C_2 - C_1 \otimes A_2, \tag{A.3}$$

$$\Delta_2 = B_1 \otimes A_2 - A_1 \otimes B_2 \tag{A.4}$$

(for details on the tensor product and relation to the MEPs, see, for example, the well known book [1]). The two-parameter eigenvalue problem (A.1) is *nonsingular* when Δ_0 is nonsingular. In this case the matrices $\Delta_0^{-1}\Delta_1$ and $\Delta_0^{-1}\Delta_2$ commute and (A.1) is equivalent to a coupled pair of GPSs

$$\begin{cases} \Delta_1\mathbf{z} = \lambda\, \Delta_0\mathbf{z}, \\ \Delta_2\mathbf{z} = \mu\, \Delta_0\mathbf{z} \end{cases} \tag{A.5}$$

for decomposable tensors $\mathbf{z} = \mathbf{x} \otimes \mathbf{y} \in S$ (see [1]).

There exist some numerical methods for two-parameter eigenvalue problems. If m is small, we can apply the existing numerical methods for the GEPs to solve the coupled pair (A.5). An algorithm of this kind, which is based on the QZ algorithm, is presented in the paper [4].

When m is large, it is not feasible to compute all eigenpairs. There are some iterative methods that can be applied to compute some solutions. Most of them require good initial approximations to avoid misconvergence. One such method, that we apply in our numerical experiments, is the (TRQI) method, from [5], which is a generalization of the standard Rayleigh quotient iteration, (see, e.g., [2]). This method computes one eigenpair at a time.

In case when we are interested in more than just one eigenpair and we do not have any initial approximations, a method of choice is a JD type method [4]. The state-of-the-art, which uses harmonic Ritz values [3], can be used to compute a small number of eigenvalues of (A.1), which are closest to a given target (λ_T, μ_T). A brief overview of the method is now presented.

For the numerical solution we exploit a JD method as developed in [3, 4, 6].

In this method the eigenvectors $\mathbf{x}$ and $\mathbf{y}$ are sought in search spaces $\mathcal{U}$ and $\mathcal{V}$, respectively. There are two main phases: expansion of the subspaces, and extraction of an approximate eigenpair from the search space. First consider the subspace expansion. Suppose that we have approximate eigenvectors $\mathbf{u} \approx \mathbf{x}$ and $\mathbf{v} \approx \mathbf{y}$ with corresponding approximate eigenvalue $(\sigma, \tau) \approx (\lambda, \mu)$; for instance, the tensor Rayleigh quotient. We are interested in orthogonal improvements $s \perp u$ and $t \perp v$ such that

$$A_1(\mathbf{u} + \mathbf{s}) = \lambda B_1(\mathbf{u} + \mathbf{s}) + \mu C_1(\mathbf{u} + \mathbf{s}), \tag{A.6}$$

$$A_2(\mathbf{v} + \mathbf{t}) = \lambda B_2(\mathbf{v} + \mathbf{t}) + \mu C_2(\mathbf{v} + \mathbf{t}). \tag{A.7}$$

Let

$$r_1 = (A_1 - \sigma B_1 - \tau C_1)\,\mathbf{u}, \tag{A.8}$$

$$r_2 = (A_2 - \sigma B_2 - \tau C_2)\,\mathbf{v} \tag{A.9}$$

be the residuals of vector $u \otimes v$ and value (σ, τ). We can rewrite (A.6) and (A.7) as

$$(A_1 - \sigma B_1 - \tau C_1)\,\mathbf{s} = -\mathbf{r}_1 + (\lambda - \sigma)B_1\mathbf{u} + (\mu - \tau)\,C_1\mathbf{u} \tag{A.10}$$
$$+ (\lambda - \sigma)\,B_1\mathbf{s} + (\mu - \tau)\,C_1\mathbf{s},$$

$$(A_2 - \sigma B_2 - \tau C_2)\,\mathbf{t} = -\mathbf{r}_2 + (\lambda - \sigma)\,B_2\mathbf{v} + (\mu - \tau)\,C_2\mathbf{v} \tag{A.11}$$
$$+ (\lambda - \sigma)\,B_2\mathbf{t} + (\mu - \tau)\,C_2\mathbf{t}.$$

We neglect the second-order correction terms $(\lambda - \sigma)\,B_1\mathbf{s} + (\mu - \tau)\,C_1\mathbf{s}$ and $(\lambda - \sigma)\,B_2\mathbf{t} + (\mu - \tau)\,C_2\mathbf{t}$. Let V be a real matrix of order $(n_1 + n_2) \times 2$ with columns (for reasons of stability, preferably orthonormal) such that

$$\text{span}(V) = \text{span}\left(\begin{bmatrix} B_1\mathbf{u} \\ B_2\mathbf{v} \end{bmatrix}, \begin{bmatrix} C_1\mathbf{u} \\ C_2\mathbf{v} \end{bmatrix}\right),$$

and let W be another real matrix of order $(n_1 + n_2) \times 2$ defined by

$$W = \begin{bmatrix} u & 0 \\ 0 & v \end{bmatrix}.$$

With the oblique projection

$$P = I - V(W^T V)^{-1} W^T$$

onto $(V)^{\perp}$ along (W), it follows that

$$Pr = \mathbf{r}, \quad P\begin{bmatrix} B_1\mathbf{u} \\ B_2\mathbf{v} \end{bmatrix} = P\begin{bmatrix} C_1\mathbf{u} \\ C_2\mathbf{v} \end{bmatrix} = 0, \tag{A.12}$$

where $r = [r_1 \ r_2]^T$. Therefore, we can project out the first-order terms $(\lambda - \sigma) B_1 \mathbf{u} + (\mu - \tau) C_1 \mathbf{u}$ and $(\lambda - \sigma) B_2 \mathbf{v} + (\mu - \tau) C_2 \mathbf{v}$ using this oblique projection, reformulating (A.10) and (A.11) (without the neglected second-order correction terms) as

$$
P \begin{bmatrix} A_1 - \sigma B_1 - \tau C_1 & 0 \\ 0 & A_2 - \sigma B_2 - \tau C_2 \end{bmatrix} \begin{bmatrix} \mathbf{s} \\ \mathbf{t} \end{bmatrix} = - \begin{bmatrix} \mathbf{r}_1 \\ \mathbf{r}_2 \end{bmatrix} \tag{A.13}
$$

for $s \perp u$ and $t \perp v$. We use (possibly) inexact solutions $\mathbf{s}$ and $\mathbf{t}$ to this linear system to expand the search spaces $\mathcal{U}$ and $\mathcal{V}$.

Now we focus on the subspace extraction. As introduced in [3], the harmonic Rayleigh–Ritz extraction for the MEP extracts approximate vectors $\mathbf{u}$, $\mathbf{v}$ and corresponding values σ and τ by imposing the Galerkin conditions

$$
\begin{aligned}
A_1 \mathbf{u} - \sigma B_1 \mathbf{u} - \tau C_1 \mathbf{u} &\perp (A_1 - \sigma B_1 - \tau C_1) \, \mathcal{U}, \\
A_2 \mathbf{v} - \sigma B_2 \mathbf{v} - \tau C_2 \mathbf{v} &\perp (A_2 - \sigma B_2 - \tau C_2) \, \mathcal{V}.
\end{aligned} \tag{A.14}
$$

This generally turns out to be a method of choice for interior eigenvalues near a target (σ, τ). A basic pseudocode for the method is given in the following Algorithm, with RGS or any other numerically robust method to expand an orthonormal basis.

Algorithm
A *JD* type method for the MEP
Input: Starting vectors $\mathbf{u}$, $\mathbf{v}$ and a tolerance ε
Output: An approximate eigenpair (θ, η, u, v) for the MEP
$\mathbf{s} = \mathbf{u}$, $\mathbf{t} = \mathbf{v}$, $U_0 = [\,]$, $V_0 = [\,]$
for $k = 1, 2, \ldots$
    $\mathrm{RGS}(U_{k-1}, \mathbf{s}) \to U_k$, $\mathrm{RGS}(V_{k-1}, \mathbf{t}) \to V_k$
    Extract an approximation $(\mathbf{u}, \mathbf{v}, \theta, \eta)$, (A.14)
    Solve (approximately) $\mathbf{s} \perp \mathbf{u}$, $\mathbf{t} \perp \mathbf{v}$, (A.13)
end

References

1. Atkinson, F.V.: Multiparameter Eigenvalue Problems. Academic Press, New York (1972)
2. Golub, G.H., Van Loan, C.F.: Matrix Computations, 3rd edn. The Johns Hopkins University Press, Baltimore (1996)
3. Hochstenbach, M.E., Plestenjak, B.: Harmonic Rayleigh-Ritz for the multiparameter eigenvalue problem. Electron. Trans. Numer. Anal. **29**, 81–96 (2008)
4. Hochstenbach, M.E., Košir, T., Plestenjak, B.: A Jacobi-Davidson type method for the non-singular two-parameter eigenvalue problem. SIAM J. Matrix Anal. Appl. **26**, 477–497 (2005)
5. Plestenjak, B.: A continuation method for a right definite two-parameter eigenvalue problem. SIAM J. Matrix Anal. Appl. **21**, 1163–1184 (2000)
6. Rommes, J.: Arnoldi and Jacobi-Davidson methods for generalized Eigenvalue problems $Ax = \lambda Bx$ with singular B. Math. Comput. **77**, 995–1015 (2008)

Index

C.-I. Gheorghiu, *Spectral Methods for Non-Standard Eigenvalue Problems*,
SpringerBriefs in Mathematics, DOI: 10.1007/978-3-319-06230-3,
© The Author(s) 2014